New Horizons in Piling

New Horizons in Piling

Development and Application of Press-In Piling

Malcolm D. Bolton

Emeritus Professor, Cambridge University, UK

Akio Kitamura

GIKEN Ltd., Kochi, Japan

Osamu Kusakabe

Emeritus Professor, Tokyo Institute of Technology, Japan

Masaaki Terashi

International Press-in Association, Tokyo, Japan

CRC Press
Taylor & Francis Group
Boca Raton London New York

CRC Press is an imprint of the
Taylor & Francis Group, an **informa** business

Cover photo: Rotary press-in piling for the emergency repair of a bridge abutment threatened by erosion, Kagoshima, Japan

CRC Press/Balkema is an imprint of the Taylor & Francis Group, an informa business

© 2021 Taylor & Francis Group, London, UK

Typeset by codeMantra

Library of Congress Cataloging-in-Publication Data Applied for
Published by: CRC Press/Balkema
                Schipholweg 107C, 2316 XC Leiden, The Netherlands
                e-mail: Pub.NL@taylorandfrancis.com
                www.routledge.com – www.taylorandfrancis.com

ISBN: 978-0-367-54652-6 (hbk)
ISBN: 978-1-003-09000-7 (ebk)

DOI: 10.1201/9781003090007
DOI: https://doi.org/10.1201/9781003090007

Contents

List of figures

List of tables

Preface

Urbanization is a global trend, which requires developing countries to improve the capacity of existing infrastructures. Developed countries are required to maintain and rehabilitate existing aged infrastructures. The recent increased frequency of earthquakes and continuing sea-level rise urges infrastructures' upgrade and resiliency. Modern society demands the minimum stagnation of public services even during upgrading the existing infrastructures. In order to accommodate these demands, earth retaining and/or water shutoff by means of embedded wall using prefabricated material is often an important component of a project, and embedded structures are a promising solution to add resiliency to the existing structures. However, difficulties encountered in the congested urban areas such as noise and vibration, limited working space, headroom restriction, adjacent construction, and underground obstacles prevented the wide use of embedded wall and structures.

Various press-in piling techniques, originally developed to eliminate the noise and vibration problems, turn out to be effective tool to cope with the other difficulties as well, but to varying degrees. The most appropriate press-in technique for a certain project should be chosen considering difficulties specific for the project. This book provides general information about various press-in techniques and then focuses on the unique walk-on-pile-type press-in piling machine called "Silent Piler". Unique because the machine uses pre-installed piles as a source of reaction force to jack in a new pile by hydraulic pressure, and the machine is compact and relatively light in weight. Unique also because the machine can walk along the top of piles already installed; thus, the construction work minimizes temporary work. These unique functions have a significant merit for piling works in congested urban environment and for piling in a limited space. Further development of a family of "Silent Piler" enables to do piling for harder ground, including gravel, rock layers, and even reinforced concrete block. This capability opens up a new area of application of piling works previously considered not possible.

This book is written as an introductory book aiming for readers who are not familiar with press-in piling, including project owners, design engineers, practical engineers as well as researchers and students. This introductory book contains five chapters.

Chapter 1 "Introduction" describes significant benefits of embedded structures over gravity-type structures, which has been clearly demonstrated by the recent tsunami disaster. The chapter touches on requirements for design and construction that modern society sets. Chapter 2 "Construction in press-in piling" starts from the description of various press-in piling techniques and introduces a brief history of invention and development of a family of Silent Piler, which is an invention of Japan. The chapter explains mechanical aspects of Silent Piler and construction procedures in detail. Chapter 3 "Innovative applications" presents classification of applications and introduces ample application examples, including projects of urban redevelopment and renovation, surface transportations: road and railways, and inland water. Chapter 4 "Emerging applications" provides further interesting applications, including projects of maritime transport infrastructure, coastal and beach protection, seismic reinforcement, and other interesting applications. Chapter 5 "Responses of piles installed by the press-in method" describes geotechnical aspects of the press-in technology, introducing responses of piles installed by press-in piling, mainly based on the research outcomes obtained from the research collaborations between the University of Cambridge and GIKEN LTD. for more than 20 years. Issues presented in this chapter include "Ground vibration and noise during pile installation", "Ground responses and pile resistance during pile installation", and "Performance of a single pile installed by press-in piling". In the Appendix, there is a list of publications related to press-in piling appeared in different journals and conferences, including the first International Conference of Press-in Engineering held in 2018.

The publication of this book would not have been possible without supports from various individuals. The authors would like to record our deepest gratitude to Messrs. Koji Kajino, Masafumi Yamaguchi, Yuta Kitano, and Hayato Nishitani and also to Ms. Nanase Ogawa for providing materials and figures.

Malcolm D. Bolton
Akio Kitamura
Osamu Kusakabe
Masaaki Terashi

Important: "SILENT PILER" is a registered trademark or trademark of GIKEN LTD. in the United States and other countries. Throughout this book, "Silent Piler" is used instead of "SILENT PILER" for readability, except Table 2.1 which lists the product names.

Chapter 1

Introduction

1.1 Embedded structures

The Great East Japan Earthquake with a moment magnitude (M) of 9.0 occurred along a subduction zone in the Pacific Ocean on March 11, 2011, and the earthquake triggered enormous tsunami disasters as shown in Figure 1.1. The maximum height of over 21 m was recorded at Fukushima Prefecture. Intensive investigation of damage to structures demonstrates significant advantages of embedded structures over gravity-type structures against tsunami disasters.

A typical gravity-type structure is a breakwater caisson constructed on a rubble mound, as illustrated in Figure 1.2.

a) Wave overtopping
(The Mainichi photobank)

b) Stranded vessel

Figure 1.1 Tsunami disasters caused by the Great East Japan Earthquake. (a) Wave overtopping (the Mainichi photobank). (b) Stranded vessel.

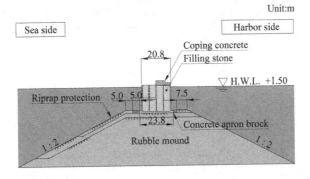

Figure 1.2 Typical breakwater caisson on a rubble mound.

Figure 1.3 is a view of the failed caisson-type breakwater constructed on a rubble mound in Kamaishi Port, of which the depth is −63 m, the Deepest Breakwater according to Guinness World Records. Twelve out of 22 caissons were damaged.

Gravity-type structures must meet three stability requirements: (i) stability of vertical bearing capacity, (ii) sliding stability, and (iii) overturning moment stability. The situation may be idealized as a footing with a width of B resting on the ground surface subjected to combined loads of vertical load (V), horizontal load (H), and moment (M) as illustrated in Figure 1.4.

Theoretical and experimental evidence on the failure of the footing can be summarized in a three-dimensional form of failure envelope shown in Figure 1.5. Working load conditions must be within the failure envelope for safe operation.

Tsunami flow reduces the effective self-weight of the structure due to the buoyancy force, which is generally considered to increase the safety against vertical bearing failure, although the bearing capacity may decrease to some extent when the tsunami

Figure 1.3 A view of failed breakwater in Kamaishi Port. (Tohoku Regional Development Bureau, MLIT Japan).

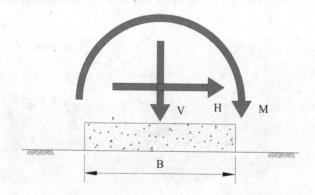

Figure 1.4 Footing subjected to combined loads.

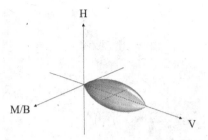

Figure 1.5 Failure envelope of a footing.

flow generates seepage in a rubble mound. The buoyancy force, in turn, decreases frictional resistance at the base of the footing, and the tsunami flow also increases the horizontal force acting on the foreside wall of the structure due to the water level difference between the sea side and the harbor side. These two combined effects would decrease the safety against sliding failure. The tsunami flow also greatly increases the overturning moment because the moment arm from the base to the point of resultant horizontal forces becomes longer, and meanwhile the horizontal force also increases in magnitude. These resultant effects lead to a decrease in safety against overturning failure. Failure modes, therefore, may vary with the loading conditions to which the structure is subjected.

When a tsunami overflows a breakwater, it causes scouring of the mound below the structures which may then tilt or topple. Hydraulic engineers also point out that negative pressure generated on the backside of the structure under a tsunami overflow causes horizontal destabilizing forces to increase based on laboratory experiments.

Another typical example of gravity-type structure is buildings with spread foundation. A number of buildings and residential houses with spread foundations were swept away due to tsunami flows, as shown in Figure 1.6.

Figure 1.6 Overturned building with spread footing.

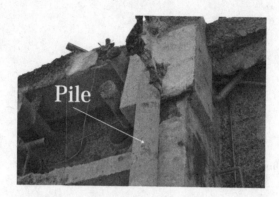

Figure 1.7 Toppled four-story building with pile foundation (Tamura, 2012).

Figure 1.7 is a toppled four-story, steel frame building supported by a friction pile foundation consisting of pre-stressed concrete piles with a diameter of 300 mm. In Figure 1.7, it is seen that one pile is hanging from its foundation. Building engineers consider that the combined effects of buoyancy forces and horizontal forces may be the main reason for the damage. Insufficient embedment may mean that piled foundations are not sufficiently resilient against strong tsunami flows. Some geotechnical engineers suspect that soil liquefaction may have occurred prior to the arrival of the tsunami flow at this particular location, reducing the pullout resistance of the friction piles.

However, properly embedded structures can be considered to be resilient against even a severe tsunami. The embedment effects increase the vertical bearing capacity largely due to overburden pressure, and increase the pullout resistance (negative vertical bearing capacity) due to frictional resistance along the periphery of the embedded parts of the structure. The embedment effects also increase the horizontal resistance mainly due to the mobilization of passive earth pressure on the side walls of the embedded elements of the structure. Similarly, the moment resistance increases benefitted by the mobilization of passive earth pressure as well as frictional resistance at the side walls and at the base of the structure.

Referring to the rugby ball-shaped failure envelope shown in Figure 1.5, the embedment effects significantly enlarge the size of the failure envelope and create additional pullout resistance, as illustrated in Figure 1.8.

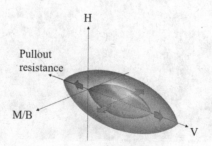

Figure 1.8 Effect of embedment on failure envelope.

When the structure is firmly embedded, and the failure envelope becomes sufficiently large relative to working loads, an unconfined failure mechanism is virtually impossible to form and the structure behaves in a resilient manner.

Figure 1.9 is a view of a temporary double-walled cofferdam comprising two parallel rows of steel sheet piles driven into the ground and connected together by a system of tie rods at one level. The space between the walls was filled with granular materials such as sand, gravel, or broken rock. Figure 1.10 illustrates the situations before and after the tsunami, indicating about 2 m depth of erosion on the sea side.

As shown in Figure 1.9, no structural damage was observed either at the steel sheet piles or of the tie rods, although the upper part of the backfill materials was scoured and washed away to about 1 m depth. This example clearly demonstrates that embedded structures are resilient and have advantages over gravity-type structures against tsunami disasters. A sufficiently embedded structure may be designated an "implant structure".

Figure 1.9 Double-walled cofferdam with no structural damage.

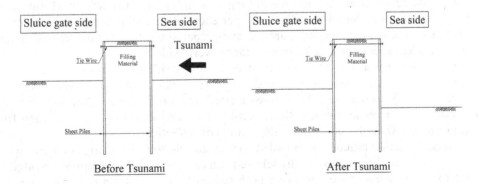

Figure 1.10 Situations before and after tsunami.

1.2 Steel as a construction material

Steel is a ductile material with high tensile strength in comparison with other major construction materials such as soil, rock, and concrete. Steel makers supply steel for a wide range of engineering applications, including civil engineering applications such as foundations, for port and harbor structures, and for forestry conservation.

Geotechnical engineers have long been familiar with U-shaped steel sheet piles for temporary retaining construction and coffering as well as permanent structures. Steel tubular piles are also widely used for civil engineering structures, such as retaining walls, foundations, port and harbor facilities, countermeasures for landslides, and more recently liquefaction countermeasures for river and coastal levees.

According to the statistics of the year 2014 published by the Japan Iron and Steel Federation, demand for steel sheet piles in Japan was as much as 597,000 tons in total, in which 361,000 tons were for permanent structures and 236,000 tons for temporary structures. Major areas of applications were forestry conservation and flood control (279,000 ton), port and harbor structures (38,000 tons), and others (44,000 tons).

Embedded structures using steel sheet piles/steel tubular piles offer resilient structures. The Japanese Association for Steel Pipe Piles and Steel Sheet Piles conducted damage investigations on structures after the Great East Japan Earthquake, covering river levees with seismic countermeasures, river revetments, road retaining structures, and cofferdams. All the structures adopted steel sheet piles. The investigation confirmed that no damage or minor damage was observed and all the structures remained sound. As demonstrated in Figure 1.9 as an example, the temporary double-walled cofferdams comprising two parallel rows of steel sheet piles survived against the Great East Japan Earthquake.

1.3 Design and construction requirements

Many developing countries with an increasing population have been experiencing urbanization at an accelerating rate. Existing urban infrastructure in these regions is insufficient to accommodate an increasing social pressure, both in terms of volume of demand and in terms of quality of demand. On the other hand, developed countries are facing the urgent need to deal with aging urban infrastructure and how to maintain, upgrade, and replace if necessary. Some urban infrastructures may not comply with updated safety and environmental regulations, and require refurbishment, reinforcement, or even demolition. Recent experiences of natural disasters in many parts of the world have urged public awareness and preparedness to natural disasters. Building a resilient society is now a common target worldwide.

Construction works in congested urban environments impose a variety of restrictions: space, time, environment, safety, and cost. Development of appropriate construction machinery is required that can work effectively under these restrictions: limited horizontal and vertical space, short construction period, noise- and vibration-free environment-friendly, safety operations, and cost-effective.

In construction practice, a so-called "optimum solution" for a particular construction project has been traditionally selected from a cluster of design options plotted in a two-dimensional space, satisfying both requirements for structural safety and for minimum cost.

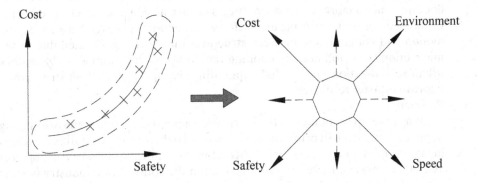

Figure 1.11 Selection of construction methods.

Modern society appreciates diverse values for a better quality of life and at the same time responding to increasing concern for the global environment, and potential disasters related to climate change associated with the greenhouse effect. The construction industry responds to those new priorities set by society. The optimum construction option should accordingly be carefully selected in view of multi-axes, representing various values, as illustrated in Figure 1.11.

Modern construction requirements may be broadly categorized into four aspects: (i) social acceptability, (ii) serviceability in operation, (iii) resilience against disaster, and (iv) productivity.

i. Social acceptability

Socially acceptable construction means construction with minimum disruption or disturbance to existing social functions and services, and on daily human activities as well as on the global environments. Such acceptability subsumes conventional optimization of structural safety, economically minimum cost, and other factors.

Noise and vibration caused by construction activities, and exceeding allowable levels, is not acceptable, in particular in congested urban environments. Minimum disruption or disturbance to existing facilities and daily services is required with a construction period as short as possible. Emissions of carbon dioxide and other wastes from construction activities must also be at a minimum.

Structures must also be compatible with the existing urban environments, in shape, color, and design.

ii. Serviceability in operation

Civil structures have been in service for a long period of time, many years and even centuries. These structures should remain in a sound state for their service period. Easy retrofitting functions and upgrading possibilities are also important for maintaining continuous services, as are esthetics considerations.

iii. Resilience against disaster

Civil structures are subjected to various natural disasters depending on the regions in which the structures are constructed. Recent years have seen unprecedented large-scale earthquakes in earthquake-prone regions. Many areas throughout the world are also suffering water disasters related to climate change, flooding, due to more severe storm surges occurring while sea level steadily rises. Combined disasters also occur, such as a combined disaster between seismic motion and tsunami, and between structures that have deteriorated due to seismic motion being followed by flooding caused by extreme rainfalls. The issues of urban revitalization must include upgrading the safety of existing structures to improve disaster resilience.

iv. Productivity

Many developed countries are experiencing a rapid reduction in the working-age population, which will bring about various social difficulties, including a decrease in tax revenues, budget cuts for infrastructure investments, and a lack of engineers and other workers. Increasing productivity in the construction industry is an urgent issue to be solved. Such difficulties may be partially eased by other aspects of modern society, which are the rapid development of technology especially in the areas of ICT (Information and Communication Technology), AI (Artificial Intelligence), and Robotics. It can be envisaged in the very near future that major construction processes proceed with unmanned construction machines. This anticipation suggests that automated construction machinery will play a significant role in future construction activities.

To meet those four requirements, novel construction machinery and technology is needed. In the field of foundation engineering, various types of press-in piling are now available to provide practical solutions. The subsequent chapter begins with the classification of these press-in piling.

Reference

Tamura, S. (2012) Building Damage Caused by Tsunami of the 2011 Great East Japan Earthquake, Annuals of Disaster Prevention Research Institute, Kyoto University, No.55A, p. 187, Photo 34.

Chapter 2

Construction by press-in piling

2.1 Classification of press-in piling methods

Embedded walls and structures are often constructed by installing prefabricated materials, such as piles and sheet piles, into the ground as displacement piles. Pile installation methods are commonly classified into three categories: impact driving, vibratory driving, and press-in piling. Traditional pile driving either by impact hammer or by vibratory hammer, which inevitably accompanies nuisance such as noise and vibration, has become unacceptable in the construction in sensitive areas.

Attempts to develop press-in piling, noise- and vibration-free piling techniques, have a relatively long history. White & Deeks (2007) presented an excellent summary of the development of piling machines for jacked foundation piles. Currently, the press-in piling can be classified into the following three sub-categories with respect to the source of reaction as well as the point of application of press-in force, as presented in Table 2.1.

The choice of the most appropriate piling technique depends not only on the direct cost for pile installation but also on various conditions/requirements as summarized in Chapter 1. It is important to know and update the information on piling techniques and their applications.

Table 2.1 Classification of press-in piling methods

Sub-category	Major source of reaction	Point of press-in force	Typical product name
Deadweight-type press-in piling	Counterweight	Pile top or Pile periphery	G-pile Modular Piling System
Panel-type press-in piling	Weight of rams & piles plus Pullout resistance of locked piles during penetration of the other piles	Pile tops	(suspended rams) Pilemaster
			(rig-mounted rams) Hydro-Press-System, Push and Pull System
Walk-on-pile-type press-in piling	Pullout resistance of previously installed piles	Pile periphery at lower level	SILENT PILER

2.1.1 Deadweight-type press-in piling

Peng (2011) presented a brief review of the historical development of press-in machine in China, dating back to the early 1950s. In the early 1970s, a manufacturer in Changsha developed a hydraulic static piling machine with 3,200 kN capacity. The first commercially available press-in piling machine by using a hydraulic pressure was manufactured in 1975. This piling machine is classified into the deadweight type, since reaction is provided by kentledge and a machine itself. In the 1980s, a new generation of the updated hydraulic static pile driver was developed and there are two categories: the top pressure type and the holding type as shown in Figure 2.1a and b.

The top pressure type provides the pressure on the top of a pile, jacking the pile into the ground, while the holding type holds the sides of the pile with the help of friction. This family of machines can provide the maximum driving capacity exceeding 10,000 kN. Peng has presented an illustration of the components of hydraulic static piling machine of the holding type, as given in Figure 2.2. This type of the piling machine has been exported to neighboring countries: Vietnam, Malaysia, Singapore, and Australia. Locally different names are given to this type of piling machine. In Australia, they call it "G-pile", in Hong Kong a "pile jacking machine" (Geotechnical engineering office, 2006), while in other countries they call the machine "Hydraulic Static Pile Driver". The mechanical principles are all originated from the Chinese machine described above.

Referring to Figure 2.2, the mechanical principles may be understood. Essentially, the whole piling system is composed of three parts: main body, support sleeper attached to the main body, and two separate parallel sleepers with guiding rails. The main body consists of four press-in cylinders, horizontal gripping part, and a platform on which a counterweight can be placed. A pile is set through the gripping part and pressed into the ground by the press-in cylinders. At a given location, the main body can move Y direction within the movable range of the two separate parallel sleepers with guiding rails, by operating hydraulic cylinders horizontally located on the parallel sleepers to pull the main body horizontally, together with the operation of the four hydraulic cylinders located on the parallel sleepers to lift the main body. The whole piling system

a) Top pressure type   b) Holding type

Figure 2.1 Hydraulic static pile driver in China (Peng, 2011). (a) Top pressure type. (b) Holding type.

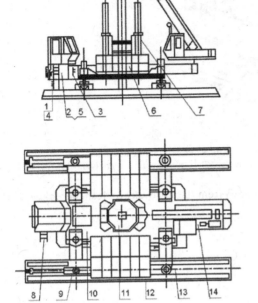

Components of full hydraulic pile driver
1 Operation cab;
2 Hydraulic assembly room;
3 Tank system;
4 Electrical system;
5 Hydraulic system;
6 Gray Iron counter weight;
7 Guiding Pile frame;
8 Ladder;
9 Pedal;
10 Leg platform structure;
11 Holding mechanism;
12 Long boat walking mechanism

Figure 2.2 Components of hydraulic static pile driver (Peng, 2011).

can also go forward like a measuring worm, by operating the four hydraulic cylinders attached to the main body to lift the parallel sleepers.

The piling machine has a gripping mechanism in the center, by which shape PC pile and cylindrical PHC (Pre-tensioned spun High strength Concrete) pile as well as H-shaped steel pile can be gripped and pressed into the ground. Lateral movement mechanism in the horizontal direction allows operators to install several piles once the whole system is fixed in location.

According to a Malaysian experience of using G-pile of 6,000 kN (Tan & Ling, 2001), the specifications are L × W × H being 11.1 × 10.0 m × 9.1 m, self-weight 200 tons, counterweight 430 tons. The main body has a dimension of 3 × 0.55 m.

A very much similar piling machine of the gripping sideway type is reported by Doubrovsky & Meshcheryakov (2015), as illustrated in Figure 2.3. The machine is called "Modular Coordinating Piling System", which equips with a hydraulic piling machine, enabling press-in and extraction, a modular coordinating skidding system designed to install piles with a flow-line production method and traverse guides. Thus, the system can provide two-axis movement of the piling machine. Specifications for the Modular Coordinating Piling System are two-axis controlled movement, 2.1 m/min of motion speed, 10 mm positional precision, and 6 piles installation per hour. The piling

Modular section of the piling system. (1) Press-in piling machine.
(2) Longitudinal guides (skid tracks).(3) Transverse guides (cross slide).

Figure 2.3 Modular Coordinating Piling System (Doubrovsky & Meshcheryakov, 2015).

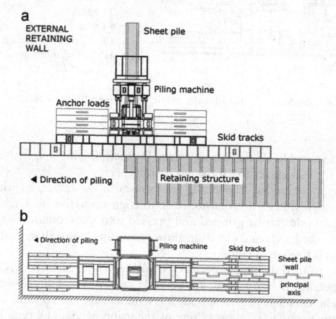

Figure 2.4 Installation of sheet pile with a flow-line method (Doubrovsky & Meshcheryakov, 2015). (a) Front view. (b) Plan view.

machine uses a side wedge clamping system for pressing in of the PC pile, sheet piles, and other construction elements with a cross-section up to 500 mm and a press-in pressure of up to 2,300 kN. Figure 2.4 illustrates the construction of sheet pile wall by the method. Piles can be installed at a distance of about 1 m from existing structures.

As this type of piling machine can obtain reaction from kentledge, there is no restriction on pile spacing between adjacent piles. Once the whole system is assembled at site, several piles can be installed only by moving the hydraulic jacks within the foot print. The system is best suited for installing building foundation piles. In order to obtain sufficient reaction, the system tends to become larger in size and heavier in weight, it is suitable for sites with fairly large and flat ground, and it is not practical for soft ground and smaller construction sites with limited head clearance.

2.1.2 Panel-type press-in piling

Piling techniques classified as panel-type press-in piling utilize the combined weight of hydraulic jacks and the weight of piles as a source of reaction in the initial phase of penetration. At a certain depth onward, a limited number of piles are pressed-in while the jacks for the rest of piles are locked and their pullout resistances provide the additional reaction for pile installation.

In the UK in the 1960s, there existed a machine called the "Pilemaster" for reducing noise and ground vibration during steel sheet pile installation by static jacking, as shown in Figure 2.5.

This particular machine was equipped with eight hydraulic jacks and was primarily designed and used for panel driving, where eight steel sheet piles are threaded together above the ground in a support frame so as to form a panel prior to jacking, to maintain the required alignment and verticality of the installed piles and the good performance of connected locks. The source of reaction forces came partly from the self-weight of the hydraulic jack assembly hung from a crane and the steel sheet piles themselves and partly from the pullout resistance of embedded steel sheet piles during the course of pile installation. Because the pile installation by the Pilemaster involved a temporary support frame for forming a panel of piles prior to jacking, and also required a crane to suspend the machine throughout the installation processes, the practicability of the Pilemaster was limited and this type of machine did not gain wide acceptance in practice.

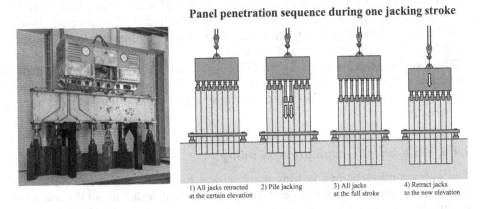

Panel penetration sequence during one jacking stroke

1) All jacks retracted at the certain elevation   2) Pile jacking   3) All jacks at the full stroke   4) Retract jacks to the new elevation

Figure 2.5 Pilemaster.

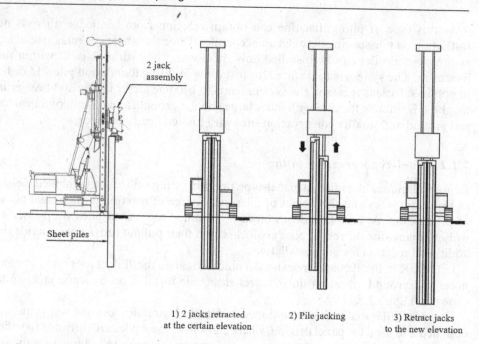

2 jack assembly

Sheet piles

1) 2 jacks retracted
at the certain elevation

2) Pile jacking

3) Retract jacks
to the new elevation

Figure 2.6 Rig-mounted panel-type pile installation machine.

In more recently developed panel-type press-in piling machines, the hydraulic jack assembly can be mounted on the leader of multi-purpose piling rig, appropriately selected by considering the stability of the rig and power supply capacity. Hydraulic jack assembly is equipped with two to six jacks, and a panel of two to six sheet piles is jacked into the ground. As the leader of piling rig provides the correct pile alignment and verticality, the temporary support frame is no more necessary. The piling sequence of rig-mounted type is similar to the crane suspended type as illustrated in Figure 2.6. The weight of the rig provides the additional reaction force for pile installation.

Different types of hydraulic jack assembly are available, and Z-shaped steel sheet pile, U-shaped steel sheet pile, and H pile profiles of varying size can be utilized depending on the type. Standard Z piles can be installed close to property boundaries. The base carrier provides rig-mounted type higher mobility than crane suspended type.

The panel-type press-in piling machines, both suspended and rig-mounted, are suitable for sheet pile wall construction. However, it is necessary to secure sufficient working space and stable working platform all along the wall for safe and accurate construction.

2.1.3 Walk-on-pile-type press-in piling

The walk-on-pile-type pile jacking machine was put into practice in 1975 in Japan, which was named the "Silent Piler". The piler can "self-walk" along the row of piles under construction without the help of crane. A new pile is pitched into the chuck of the piler and statically pressed into the ground while gripping the top of previously installed piles and gaining reaction force from their pullout resistance as shown later in Figures 2.12 and 2.13.

Silent Piler is a compact machine, workable in a limited space and head clearance with self-walking function and may be more suitable press-in piling in urban construction environments. When used with clamp crane and pile transporter, all the piling work can be completed on the tops of pile wall and eliminate temporary works. Further details of the Silent Piler will be discussed in the following sections.

2.2 Invention and development of Silent Piler

2.2.1 Introduction

Necessity is the mother of invention. Invention, however, does not emerge from necessity alone. Invention can only emerge through a person with rich creativity and solid determination. Invention of the "Silent Piler" is no exception.

Akio Kitamura, the inventor of the Silent Piler, started his business in the construction industry in 1967 with the idea of its being an anti-pollution company. During the 1960s to the early 1970s when Japan was enjoying high economic growth, Kitamura, who had ample experiences of handling a variety of construction machinery, was running his business mainly with piling contracts for foundation works using construction machinery such as the vibratory hammer.

One day, an incident occurred where an operator in his company was chased by local residents near the construction site due to complaints of noise and ground vibration caused by a vibratory hammer. He came to encounter the frequent occurrence of damage caused by pile driving works, such as roof tiles slipping down, doors unable to open, and walls cracking. He then realized that he himself had become a source of pollution in spite of his ambition to develop an anti-pollution company. He made up his mind to devote himself to inventing and developing pollution-free piling machines.

Luckily enough, he became acquainted with an experienced mechanical engineer, Yasuo Kakiuchi, as a working partner for design and manufacture. The first prototype of the Silent Piler was completed in 1975, as a piling machine offering a noise- and ground vibration-free machine. This was the point of departure for the development of the walk-on-pile-type press-in piling.

2.2.2 From original concept to design, manufacture, and selling of the Silent Piler

When he started developing the ideas, a scene recurred in his mind from, when he was watching with keen interest an operation to extract an embedded H-shaped steel pile (hereinafter called H pile) used for retaining walls in a building project. Many workers on the construction site were trying hard to pull up the H pile with a manual pulling system using timbers installed around the H pile as the required reaction. The resistance of the H pile was so strong that the pile could not easily be pulled out.

While he was recapturing the past scene, an idea flashed in his mind. A pile could be statically jacked in (hereafter "pressed-in" is used, with the same meaning as "jacked in") without noise and ground vibration, if the piling machine were designed to firmly grip the tops of existing piles previously installed in the ground, utilizing the pullout resistances of these piles as a reaction force. A clear image of piles gripping the Earth occurred to his mind.

Figure 2.7 presents a conceptual sketch of such a piling machine, where the piling machine was mounted on the top of installed piles while statically pressing a new pile into the ground by a hydraulic pressure.

Design of the first prototype commenced in 1973, and after 2 years of development, the first prototype was completed in 1975. The prototype machine had a capability of jacking a 400 mm width U-shaped steel sheet pile into the ground using hydraulic pressures. The machine was named the "Silent Piler", implying that the piling machine could install a pile in the ground quietly. Figure 2.8 shows a view of the first prototype, and Figure 2.9 presents the corresponding design drawings.

The first prototype had a designed maximum press-in force of 100 tons and the weight of the machine was 13 tons (a ratio of maximum press-in force to machine weight of 7.7). Based on a number of trial-and-error experiences on the first prototype, Kitamura and his partner improved the original design and manufactured the second prototype in 1976, as shown in Figure 2.10.

The second prototype had a designed maximum press-in force of 120 tons and a machine weight of 6 tons (a ratio of maximum press-in force to machine weight of 20), achieving high-operating performance with light weight. The second prototype had a capability of pressing-in a 6 m long steel sheet pile at a penetration rate of 0.8~2.5 m/min.

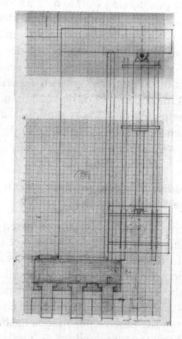

Figure 2.7 A conceptual sketch by Akio Kitamura.

Figure 2.8 First Silent Piler.

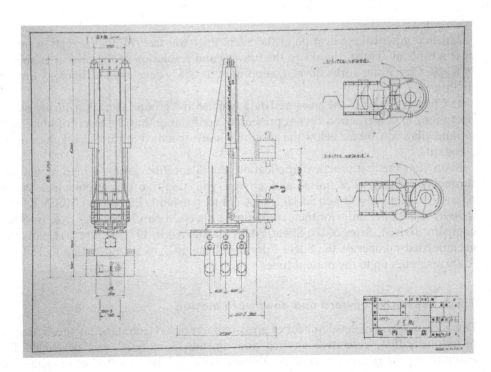

Figure 2.9 Design drawing of the first Silent Piler.

Figure 2.10 Second prototype.

The first practical application of the Silent Piler was in 1976, for excavation works near a residential area, thanks to the bravery and resolution of the project engineer, in Kochi City Waterworks Bureau, adopting this newly developed technology for the first time.

In Chapter 5, a set of the measured data of noise and ground vibration during piling operations at construction sites is presented, confirming that the levels of noise and ground vibration are far below the allowable levels specified in various regulations or standards.

Soon after the first practical application, the Silent Piler gained a high reputation throughout Japan as a noise- and ground vibration-free piling machine. While Kitamura's company, Giken Seisakusho Co. Ltd. (currently known as GIKEN LTD.), used its Silent Piler units for their own piling projects, a purchase order from another company arrived. Sales of the Silent Piler then started in 1977. Figure 2.11 plots the accumulated number of the Silent Piler sold from 1977 to 2018, totaling more than 3,400 machines up to the present time.

2.2.3 Repetitive upward and downward motion

During the course of development of press-in operational technique on-site, a unique technique called "repetitive upward and downward motion" was developed for maintaining the required alignment and verticality of the pile and for reducing the penetration resistance by repeating up and down motion of an installing pile controlled

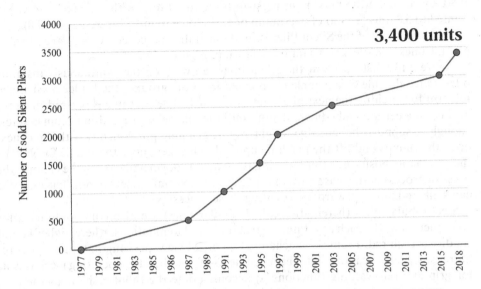

Figure 2.11 Accumulated number of sold units of Silent Piler from 1977 to 2018.

by operating a hydraulic pressure in the set of cylinders mounted on the Silent Piler. The repetitive upward and downward motion further provides the merits such as adjustment of distortion and rotation of piles and prevention of interlock separation.

The repetitive upward and downward motion is characterized by penetration length lp (downward motion) and extraction length le (upward motion), practically penetration length lp is equal or longer than extraction length le.

The upward and downward motion turns to be one of the important on-site operational techniques with respect to quality control of piles installed, thus structures to be completed. Evidence will be presented of the effectiveness of the upward and downward motion in reducing the penetration resistance, and geotechnical interpretations on the upward and downward motion will be described in Chapter 5.

2.2.4 Self-walking function

The concept of the Silent Piler stems from the idea of using the installed piles as a source of reaction force, with the machine mounted on the top of the previously installed piles which it grips. The machine then grips the next pile to be installed and statically presses the pile into the ground. This function of "Grip" forms one of the two key functions in Silent Piler.

Another key function is the "Self-Walking" function, which significantly widens the application ranges of press-in piling works. The self-walking machine means that a crane is no longer required for relocating and positioning the machine to install the next pile.

The idea of the self-walking function was already in Kitamura's mind since the design of the first prototype. To achieve this function, reducing the machine weight was essential. A unit of Silent Piler with the self-walking function was completed in

1980, by reducing the maximum press-in force to 80 tons with the machine weight limited to 5.8 tons (a ratio of maximum press-in force to machine weight of 13.8).

The main body of the Silent Piler is fundamentally composed of four parts: Leader Mast, Chuck, Saddle, and Clamps as shown in Figure 2.12.

Figure 2.13 illustrates how the Silent Piler can walk on the continuously installed piles. Once the pile ⑤ is completely pressed-in to the ground, the leader mast is slid forward for pitching the next pile ⑥ into the chuck. The next pile is pressed-in to the ground to a certain embedment depth, until the pile being pressed-in becomes stable enough to support the machine weight. While gripping pile ⑥ firmly with the chuck, open the clamps and lift the machine upward. And then move the saddle forward to the new clamp position. After lowering the machine down and closing the clamps, the press-in process of the pile ⑥ proceeds to the designed embedment depth. The machine can walk and move forward by repeating these processes.

Steel tubular piles (hereinafter called tubular piles) are also commonly used for construction works such as retaining structures and foundations, where higher lateral rigidity is required. From the mechanical points of view, pressing-in tubular piles requires some modifications onto the machine itself as well as the operating mechanisms for grip and self-walking functions to accommodate the tubular shape of the pile, while achieving a larger penetration force.

For tubular piles, two modifications are required. One is that the machine has cylindrical feet attached to the saddle, instead of clamps. The cylindrical feet are made to fit to the inner diameter of tubular pile and have a capability of expanding when firmly holding the tubular pile and of contracting when detaching from the pile. The other is the use of a driving attachment for holding the main body to move forward as illustrated in Figure 2.14.

The biggest advantage of the self-walking function is to simplify the construction procedures, that is to say, to eliminate the process of crane operation for relocating and positioning the piling machine for the next pile installation. The self-walking

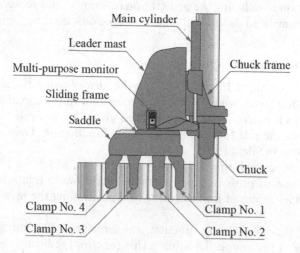

Figure 2.12 Main components of the Silent Piler.

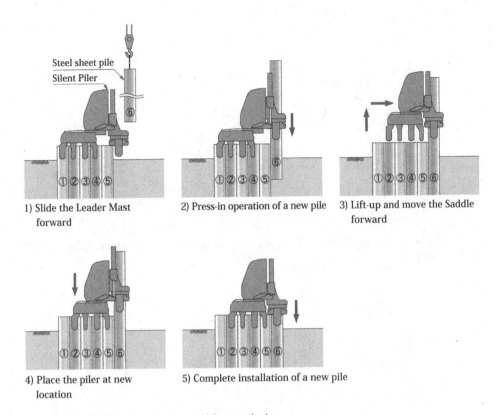

1) Slide the Leader Mast forward

2) Press-in operation of a new pile

3) Lift-up and move the Saddle forward

4) Place the piler at new location

5) Complete installation of a new pile

Figure 2.13 Self-walking function (sheet piles).

function is vital, in particular, for construction projects where only a very narrow working space is available, for example, piling works in densely populated residential areas as well as in areas near busy roads or where headroom clearance is limited, such as piling works under existing bridges.

Figure 2.15 presents various examples of piling works carried out in a narrow space, including a case where horizontal space is just over the machine width, while Figure 2.16 shows a view of piling works under an existing bridge without disturbing active traffic, where the headroom clearance was as low as 1.0 m.

These "Grip" and "Self-Walking" functions have further extended to the idea of the so-called the Non-staging system shown in Figure 2.17, where all the necessary piling work procedures, including pile transportation, pile pitching, and press-in operations can be carried out on the top of the aligned installed piles. This means that piling works can be carried out without temporary structures or temporary works, which has significant implications for widening the applications of press-in technology.

Figure 2.17 illustrates the typical configuration of the Non-staging system, also known as GRB (Giken Reaction Base) system, consisting of Silent Piler, piler stage, power unit, unit runner, clamping crane, and pile runner. Figure 2.18 shows several examples of Non-staging piling works. Chapters 3 and 4 compile case histories of

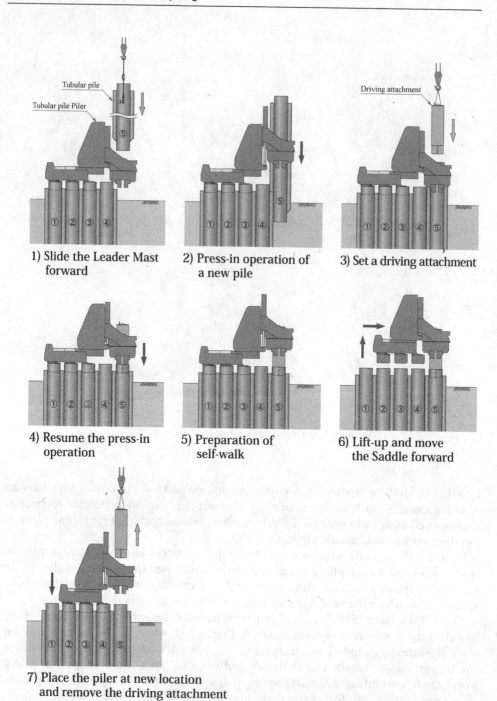

Figure 2.14 Self-walking function (for tubular pile with interlock).

Figure 2.15 Pile installation in narrow working space.

Figure 2.16 Pile installation in a narrow working space.

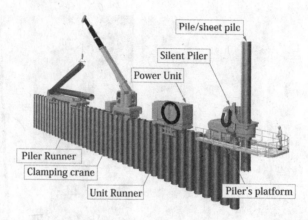

Figure 2.17 Non-staging system.

Figure 2.18 Examples of Non-staging piling works.

piling works where the piling projects could not be carried out without using these unique features of the Silent Piler.

2.2.5 Development of rotary press-in piling

A rotary pressing-in mechanism was added to the family of Silent Piler. A machine for the special purposes of pressing-in tubular piles was designed and completed in 1988

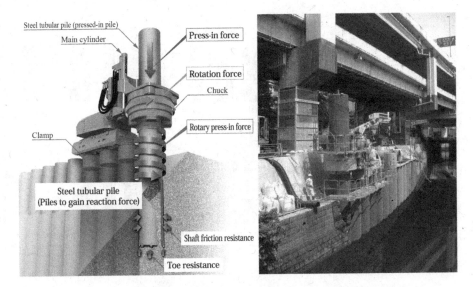

Figure 2.19 Gyro Piler and a view of operation.

after 2 years' development efforts. Further development has led to the evolution of the so-called "Gyro Piler", by which a tubular pile with cutting bits attached to the pile toe can be rotary-pressed into the ground, even in rock or in the locations having obstacles such as existing reinforced concrete structures. Practical experiences have proved that the rotary press-in piling can be applied to ground with an equivalent SPT (Standard Penetration Test) N value as high as 1,500. Figure 2.19 shows a view of the Gyro Piler and the associated operation sequences.

2.2.6 Expanding the range of applications

Expanding applications demand a wide variety of prefabricated piles in terms of shape and flexural rigidity according to the geometries and loading conditions encountered. Silent Piler is able to install most of the prefabricated piles available on the global market as shown in Figure 2.20. Figure 2.21 shows the range of section modulus and moment of inertia of prefabricated sheet piles, tubular piles, and tubular piles with P-T interlock against mass of pile per unit area of wall.

Among the variations shown in Figure 2.20, the "steel tubular pile with interlock" was developed in Japan to construct an interlocking tubular pile wall and is called "steel pipe sheet pile". In the press-in operation, tubular piles with pile diameters of 600–2,500 mm can be used for constructing interlocking tubular pile wall. The interlocks come in various shapes. A typical cross-section of "tubular pile with interlock" (with a diameter of 800 mm) and typical interlock shapes are shown in Figure 2.22.

Ground conditions vary from one place to another. It often happens that some driving assistances are required to install piles in stiffer ground conditions, in particular, where gravels, boulders, or rock layers are encountered.

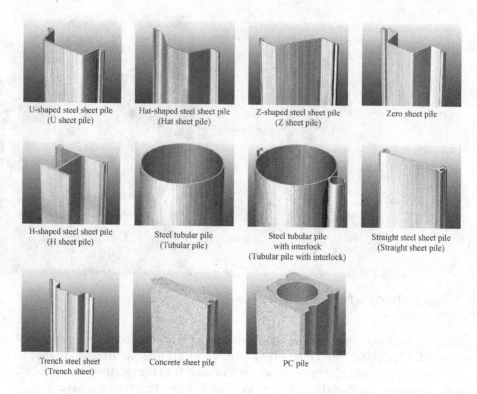

Figure 2.20 Various prefabricated piles.

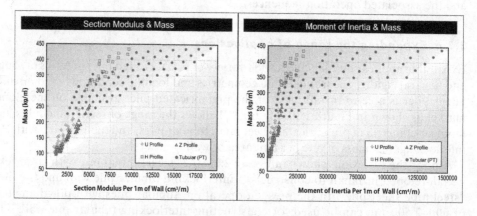

Figure 2.21 Properties of various prefabricated piles.

Although it depends on the shape, stiffness, and material of the pile to be installed, ordinary press-in operation without any driving assistance may be adopted for ground with an SPT N value less than 25. For ground with an SPT N value more than 25, certain driving assistance is often required.

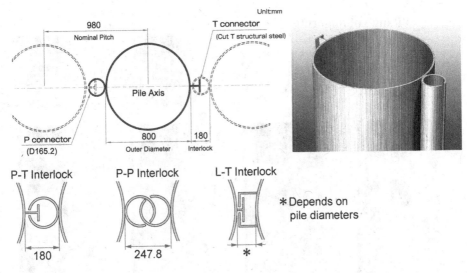

Figure 2.22 Typical cross-section of steel tubular pile with interlock.

Whenever required, the current practice of press-in operation selects one of the appropriate driving assistances, shown in Figures 2.23–2.25, depending on the ground conditions: (i) combined use of water jetting (Figure 2.23) for ground with an SPT N value typically less than 50, (ii) combined use of simultaneous inner augering (Figure 2.24) for ground with an SPT N value up to 75, and (iii) combined use of pre-boring (Figure 2.25) for ground with an SPT N value more than 75.

Penetration processes create a zone of intensified stress, often called a pressure bulb, beneath and around the pile toe. Penetration resistance increases as a zone of the pressure bulb is created. An appropriate driving assistance is selected to avoid or minimize the formation of the pressure bulb.

In the press-in with water jetting, water is supplied from a reel system atop the Silent Piler as shown in Figure 2.23. The pressurized water blasts a small diameter of water jet at the toe of sheet pile. From the principle of effective stress used in soil

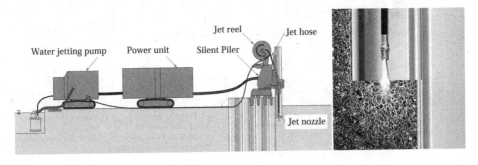

Figure 2.23 Driving assistance: combined use of water jetting.

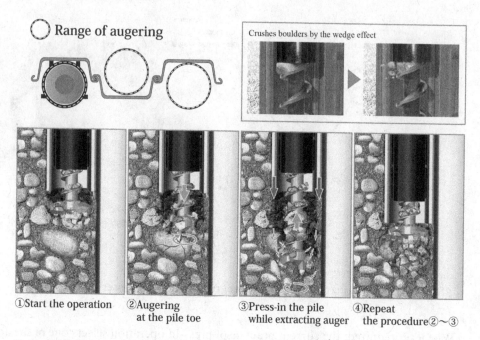

Figure 2.24 Driving assistance: combined use of inner augering.

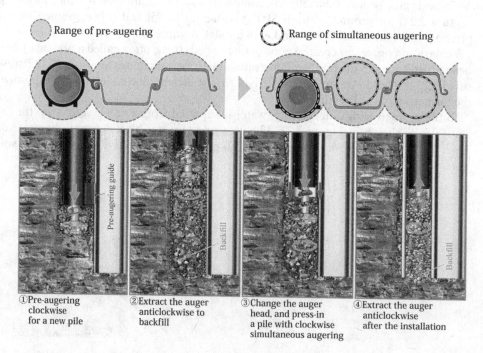

Figure 2.25 Driving assistance: combined use of pre-boring.

mechanics, it is considered that the water pressure generates an increase in pore water pressures associated with a decrease in effective stresses around the toe of the sheet pile, allowing the sheet pile to penetrate. The upward water flow generated along the interface between the sheet pile and the surrounding soil will also spread the effect of reduced effective stresses, reduce skin friction between the soil and the pile, and similarly allow easier penetration of pile into the ground. Evidence also suggests that the high-pressure water stream can punch a pilot hole through hard strata, allowing the pressed-in pile to break through, and evokes particle separation, moving smaller soil particles and allowing particles such as gravel to be pushed out of the way by the pile.

Penetration processes create a zone of a pressure bulb, beneath and around the pile toe. These penetration stresses represent the bearing capacity of the soil, which is generally expressed as a factor multiplying the effective overburden pressure at the depth of penetration. As the pile advances, the pressure bulb advances with it, and there is an outward displacement of soil similar to the process of cavity expansion. In the simultaneous inner augering depicted in Figure 2.24, an augering attachment removes soil from inside the pan of the sheet pile. The pile is then pressed-in while the auger is simultaneously extracted. The diameter of augering is smaller than the diameter of the cross-sectional area of the sheet pile to be installed, minimizing the amount of soil materials to be excavated as well as the tendency for cavity expansion. As a result, the local confining pressures are much reduced, as is the bearing capacity of the pile. It should be noted here that this simultaneous inner augering differs from conventional pre-auguring, in which the diameter of the augering zone is larger than that of the cross-sectional area of the pile to be installed and is advanced to the designed depth of the pile prior to pile installation.

For grounds with an SPT N value more than 75, the combined use of pre-boring is usually adopted. The construction works are carried out in two stages: the pre-augering, followed by the simultaneous inner augering. The ground ahead of the pile toe is pre-augered by the augering bit which is larger in cross-section than the cross-sectional area of sheet pile. When extracting the auger, the auger is set to a reverse revolution mode. This process leaves a column of loosened soil materials behind. Then the auger head is replaced with smaller one and the operation is switched to the simultaneous inner augering described above.

Table 2.2 shows a general guideline for selecting an appropriate machine and driving assistance, considering the kinds and types of pile together with the ground conditions.

2.2.7 Radio control and data collection during construction

Any construction machinery requires high safety, easy controllability, and good operability, which should result in a high quality of completed structures. The Silent Piler has acquired a radio control function from the initial stage of development.

Figure 2.26 illustrates the mechanical parts of the Silent Piler. For successful piling operations using Silent Piler, there are three fundamental control items, which are (i) leader mast control, (ii) clamp control, and (iii) chuck control. The verticality of the leader mast is crucial to press-in piles within an allowable verticality limit. The clamp control includes the opening and closing operations of the clamp for self-walking processes. The chuck holds the pile that is to be installed and controls the upward and

Table 2.2 Selection chart for appropriate Silent Piler

Group (left spanning label): **Steel sheet piles**

Piles	Types/profiles	Effective width (mm) & type	Silent Piler — Variations	Representative machine model	Standard press-in — Target ground sand, clay	Water jetting — Target ground sand, clay	Augering — Target ground sand, clay, gravel, rock	Rotary press-in piling — Target ground sand, clay, gravel, rock
U-shaped steel sheet pile (U sheet pile)	Type II–VI$_L$, Type AU, PU, GU, Type LARSSEN	400–750	U Piler	F101, F111, F201A, F301-700, F401-1400	○ (N ≤ 25)	○ (N ≤ 50)	○ (N ≤ 600)※	
Z-shaped steel sheet pile (Z sheet pile)	Type AZ, Type PZC	575–708	Z Piler	F301-700 (for single piles), F401-1400 (for double piles)	○ (N ≤ 25)	○ (N ≤ 50)	○ (N ≤ 600)※	
Hat-shaped steel sheet pile (Hat sheet pile)	10H, 25H, 45H, 50H	900	Hat Piler	F301-900	○ (N ≤ 25)	○ (N ≤ 50)	○ (N ≤ 180)※	
Zero sheet pile	NS-SP-J	600	Zero Piler	JZ100A, SCZ-ECO600S	○ (N ≤ 20)	○ (N ≤ 50)	○ (N ≤ 180)※	
H-shaped steel sheet pile (H sheet pile)	J-Domeru, NS-BOX	w: 500, H: 350–588	H-steel Piler	HP150	○ (N ≤ 7)	○ (N ≤ 30)	○	
Straight steel sheet pile (Straight sheet pile)	JFESP-FLJ, NS-SP-FL, NS-SP-FXL	500	Straight Piler	CLH150, CP50	○ (N ≤ 10)	○ (N ≤ 20)	○	
Trench steel sheet (Trench sheet)	LSP-3A, 3B, 3D, NL-3,	333	Trench Piler	ST30	○ (N ≤ 20)	○ (N ≤ 30)	○	

(Continued)

Table 2.2 (Continued) Selection chart for appropriate Silent Piler

Piles	Types/profiles	Silent Piler		Applicable ground conditions with/without a driving assistance			
	Effective width (mm) & type	Variations	Representative machine model	Standard press-in	Water jetting	Augering	Rotary press-in piling
				Target ground sand, clay	Target ground sand, clay	Target ground sand, clay, gravel, rock	Target ground sand, clay, gravel, rock
Steel tubular pile — Steel tubular pile with interlock (Tubular pile with interlock)	Dia.: 600–1,500, 2,000, 2,500 (Interlock: P-P, P-T, L-T)	Tubular pile Piler	F301-G1000 F401-G1200 F401-P1200 F501-G1200	○ (N≤15)	○ (N≤75)*		
Steel tubular pile (Tubular pile)	Dia.: 600–1,500, 2,000, 2,500	Gyro Piler	GRV2540	○	○		○

Notes:
① Extrapolated N-value applied for N values greater than 50. Estimated value for penetration depth of 30 cm, in the case that the cumulative penetration depth is less than 30 cm after 50 blows.
Extrapolated N value = N value $(50) \times 30$ (cm)/penetration depth after 50 blows (cm).
② Silent Piler models for concrete sheet piles, PC piles, and combined pile walls are also available. Further information including the Press-in-System may be found in *Press-in Retaining Structures: A Handbook*, First edition, 2016. (IPA, 2016).

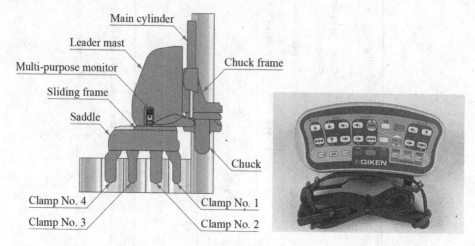

Figure 2.26 Mechanical parts of Silent Piler and a portable control panel.

downward motion of the pile during the press-in process. An operator can control these items by operating a portable control panel.

It is possible to perform press-in piling works in a semi-automatic operation mode by inputting the required data into the portable control panel, such as maximum press-in pressure applied, required penetration rate, and distance of the upward and downward motion. Full-automatic operation can be envisaged in the near future when the machine itself gathers and interprets various data during piling operations and applies a self-adjustment function using these data.

The data gathered during the piling operation are utilized for construction management and quality control. The data are displayed on a panel attached to the machine. The data include the verticality data of leader mast, the data of hydraulic pressure for clamp operation, the data of hydraulic pressure and oil flow rate for chuck operation, and the penetration depth of the pile being installed. When a driving assistance is applied, corresponding additional control data are also displayed and utilized. Figure 2.27 presents an example of the data obtained during piling operations. Utilizing a wireless information technology system, the data can be displayed in a remote office.

2.2.8 Documentation of press-in piling practice

As shown in Table 2.2, it is now possible to select an appropriate press-in machine according to the combination of the ground condition and the kind of pile to be installed. Ground conditions, in reality, exhibit a considerable variety. We must bear in mind that the press-in piling process greatly influences the quality of the completed structures.

A question arises here how to accurately install a series of piles to meet the designed requirements. We must be aware that it is not the design engineer who will install these piles but an operator. It is the experience and skill of the operator that controls

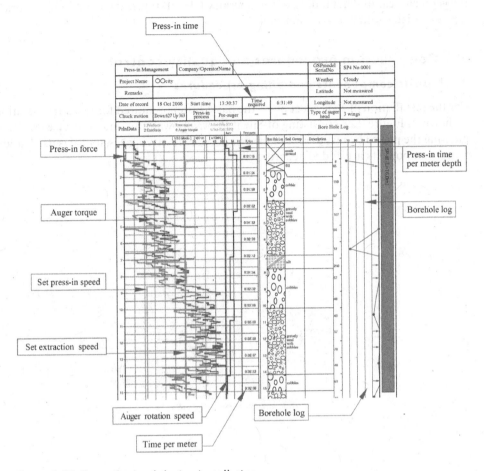

Figure 2.27 Data obtained during installation.

the efficiency and accuracy of construction, which in turn affects the performance of completed structures.

Machine–pile–soil interaction is a highly complex phenomenon and is not yet fully understood from the engineering points of view. Together with continuing efforts to promote a scientific understanding of the interaction, it is also important to gather, epitomize, and systematize the accumulated experiences in machine operation.

Based on more than 40 years experiences in piling works using the Silent Piler, a set of key operational techniques have been systematized and documented in a form of booklet in 2000. This booklet includes various aspects pertaining to press-in operation, including a unique technique of the repetitive upward and downward motion for maintaining the required alignment and verticality of the pile, while reducing the penetration resistance, and a practical technique for correcting forward and lateral lean of the pile during piling operation, whenever it occurs. The booklet serves as a learning material for inexperienced operators of the Silent Piler. Chapter 5 describes the geotechnical understanding of the operational techniques deduced from the practical experiences on construction sites.

2.3 Construction procedures and quality control

2.3.1 Basic components and piling procedures

For the installation of steel sheet piles, the basic system required for press-in operation is shown in Figure 2.28 consisting of Silent Piler, a power unit, and a service crane for pitching the pile to be installed. For cases where some driving assistance is used, additional components are added accordingly.

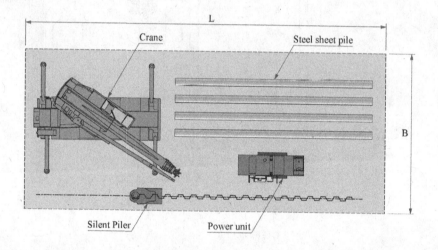

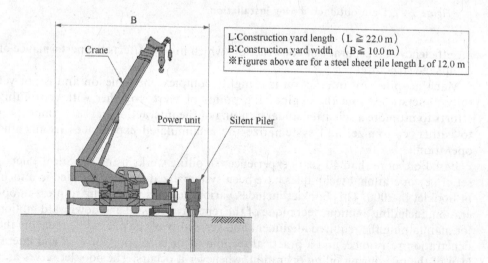

L: Construction yard length （L ≧ 22.0 m）
B: Construction yard width （B ≧ 10.0 m）
※Figures above are for a steel sheet pile length L of 12.0 m

Figure 2.28 Machine layouts of the ordinary press-in operation.

At the beginning of press-in work, there exists no pile in the ground. Therefore, an initial piling process is needed, followed by subsequent steady piling process. The procedures of the initial piling of sheet pile are as illustrated in Figures 2.29 and 2.30.

The Silent Piler and a reaction stand are placed on the ground and counterweights are loaded onto the reaction stand. In practice, sheet piles to be installed are conveniently used for the counterweight. The first sheet pile is pitched into the chuck using the service crane. The sheet pile is then pressed into the ground by lowering

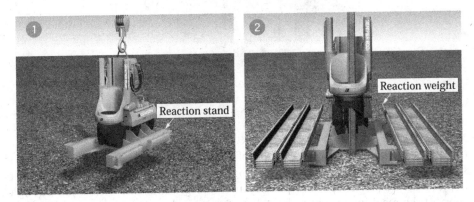

Figure 2.29 Reaction stand for initial piling work.

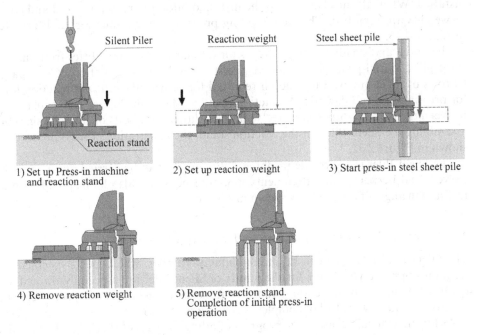

Figure 2.30 Initial press-in operation of sheet pile.

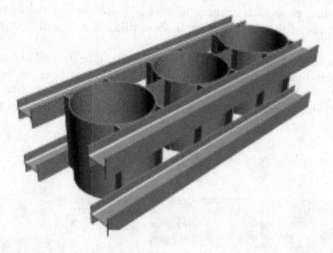

Figure 2.31 Reaction stand.

the chuck, utilizing the total weights of the machine and the counterweights as a source of reaction force. After installing the first sheet pile, the second sheet pile is pitched into the chuck and pressed into the ground until the depth enough to support the machine weight to be able to function the self-walking. By gripping the second pile, the machine goes forward and then continues to press-in the second pile to the predetermined depth. In the same manner, the third and fourth piles are installed. When all the clamps grip the installed piles, the reaction stand and counterweights are removed. The steady piling process then commences, as illustrated in Figure 2.13.

The initial and steady piling processes for tubular piles slightly differ from that for sheet piles. Initial process for tubular pile starts with the installation of two parallel rows of sheet pile wall to a depth required for providing sufficient reaction force for pressing-in a tubular pile. A reaction stand as shown in Figure 2.31 is placed in between them and tightly connected with them either with bolts or by welding. After assembling the Silent Piler on the reaction stand, the penetration process proceeds as shown in Figure 2.32.

The Silent Piler is a versatile machine. Piling works can be carried out in straight, curved, and perpendicular alignments. Battered piles can also be installed with an inclination angle of up to 30°, as shown in Figure 2.33.

2.3.2 Construction management and quality control

The Silent Piler is used for installing manufactured piles, which meets specified criteria on shipment. Operators for the family of Silent Piler are especially trained and qualified. Items in construction management basically include two aspects: schedule control and quality control of completed structures.

As for the schedule control, the progress of piling works is evaluated and controlled by comparing the actual schedule with the planned schedule. From operating

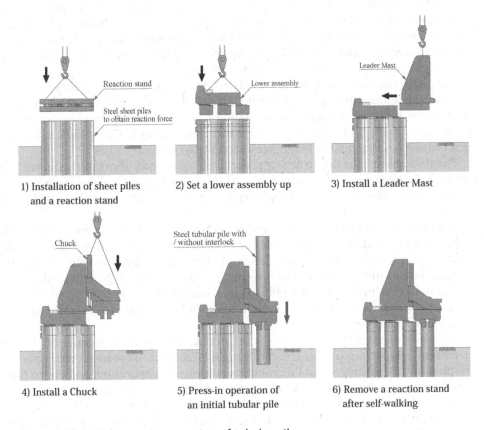

1) Installation of sheet piles and a reaction stand

2) Set a lower assembly up

3) Install a Leader Mast

4) Install a Chuck

5) Press-in operation of an initial tubular pile

6) Remove a reaction stand after self-walking

Figure 2.32 Initial press-in operation of tubular pile.

Figure 2.33 Tilting mechanism of Silent Piler.

experiences and the data in the initial part of piling works, the time required for pressing-in a pile to a designed embedment depth can normally be estimated. If any delay is noticed, modification can be introduced by possible changes of press-in operational procedures, and by considering possible combined uses of driving assistance, or by mobilizing additional number of press-in machines.

Regarding the quality control of completed structures, when the quality assurance of vertical bearing capacity of tubular piles is required, driving control also forms an item in construction management in relation to termination criteria, evidencing an estimated capacity and achieving a design embedment length. Construction management ensures overall quality assurance of completed retaining walls.

The data during the construction procedures mentioned above are gathered and collectively managed as a press-in management system. The data are analyzed, stored, and displayed in the format shown in Figure 2.27, which are submitted to the client.

The data are utilized to interpret and reconfirm the estimated ground profile with depth and the final depth of the pile toe. The data are also used to reconfirm the safety and economy of the original design. If any discrepancy is noticed between the ground profile estimated by the site investigation, and the actual ground profile based on the data obtained during the piling works, consideration must be given to possible changes, considering whether or not the presently adopted design, or the press-in procedure, should be changed.

A few comments may be appropriate on productivity. Current practice of press-in operation in Japan requires a crew of three to five technical workers: a crane operator for pitching, a qualified Silent Piler operator for press-in operation, and one to three workers assisting overall operation processes. Silent Piler contributes to shift piling projects from conventional labor-intensive work to higher productivity work.

The Ministry of Land, Infrastructures, Transport and Tourism, Japan, issues the list of daily productivity data for construction works. Although it depends on the type of sheet pile to be installed, provided that daily standard working hours is 6.3 hours, 13–15 steel sheet piles with the embedded length in the range of 15–19 m are normally installed for the ground with the maximum SPT N value less than 25. When water jetting is used as a driving assistance, 11–13 steel sheet piles must be installed for the ground with the maximum SPT N value less than 50. When pre-augering is used as a driving assistance, two to three sheet piles must be installed into the ground with SPT N value in the range between 375 and 600.

References

Doubrovsky, M.P. and Meshcheryakov, G.N. (2015) Physical modeling of sheet piles behavior to improve their numerical modeling and design, *Soils and Foundations*, Vol. 55, Issue 4, pp. 691–702.

Geotechnical Engineering Office (2006) *Foundation Design and Construction*, 66 p. The Government of the Hong Kong, Special Administrative Region.

IPA (2016) *Press-in Retaining Structures: A Handbook*, first edition. International Press-in Association (revised version will appear in 2020).

Peng, Fang-Le (2011) Current status and future prospects for press-in technology in China, *Proceedings of 3rd IPA International Workshop in China*, pp. 1–9.

Tan, S.M. and Ling C.H. (2001) The use of high capacity hydraulic injection piles for buildings in limestone ex-tin mining sites in Kuala Lumpur, *Proceedings of 14th Southeast Asian Geotechnical Conference*, Hong Kong, pp. 459–464.

White, D.J. and Deeks, A.D. (2007) Recent research into the behavior of jacked foundation piles, *Advances in Deep Foundations*, pp. 3–26.

Chapter 3

Innovative applications

3.1 Introduction

3.1.1 Increasing popularity of embedded walls and structures

Embedded walls maintain stability mostly by the vertical and horizontal resistance gained from surrounding soils at embedment. Additional resistance may be gained by tie-back or anchors if necessary. Figure 3.1 demonstrates the advantage of embedded walls by comparing the construction sequences of an embedded cantilever wall and a gravity wall for excavation support in the close vicinity of existing building. In a situation as shown in the figure, the construction of a gravity wall involves land acquisition, demolition or relocation of existing buildings, temporary earth retaining work, excavation, construction of gravity wall, and backfilling. Whereas the construction of the cantilever wall is much simpler, eliminating the adverse influence on nearby structures and resulting in reduced construction cost and construction time.

It should be noted that embedded structures are much more resilient structures than gravity-type structures when extraordinary loadings exert, for example, in tsunami disasters as described earlier in Chapter 1.

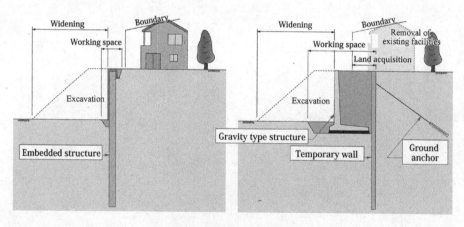

Figure 3.1 Comparison of embedded wall and gravity wall for excavation support (IPA, 2016).

For the successful construction of such an embedded wall, it is necessary to install the prefabricated piles with the required accuracy regardless of the soil type and construction circumstances. The use of traditional pile installation techniques by means of impact or vibro-hammer is often prohibited in urban areas due to noise and vibration problems. Various types of press-in piling as shown earlier in Table 2.1 succeeded in solving noise and vibration problems. However, some of the machines lack mobility and their use is limited by spatial constraints and/or the needs of stable working platform.

As already discussed in Chapter 2, the Silent Piler which obtains reaction force by gripping the previously installed piles enabled practically noise- and vibration-free pile installation. With the development of various driving assistance technologies such as jetting, simultaneous inner augering, pre-boring and rotary cutting, the Silent Piler guarantees smooth pile installation in stiffer soils and even where underground concrete structures have been left in place. Various Press-in-Systems developed on the basis of "Grip" and "Self-Walking" functions minimize the difficulties associated with construction site conditions. The Press-in-System refers to machines specially devised to cope with spatial construction constraints on-site, such as a narrow working space, headroom restrictions, and construction in the close proximity of adjacent structures. Consequently, the range of application of embedded walls and structures in civil and architectural projects has been expanded dramatically.

3.1.2 Classification of applications

There are two aspects in what we can learn from the previous applications of embedded walls and structures. One is the capability of the press-in piling machines in embedded wall construction. Specialized contractors may be most interested in this aspect, because they are responsible for accurate and efficient pile installation under given site-specific conditions (such as soil stratification and strength, spatial construction constraints, environmental regulations, influence on nearby structures, time constraints, and cost). Design engineers must have at least a minimum of necessary knowledge on this aspect in order to select functionally appropriate and constructible structures under given circumstance. Table 2.2 in Chapter 2 summarizes the information on this aspect of capability. The table starts with the type and size of piles and corresponding type of Silent Piler. For each combination, the table shows the maximum soil strength in terms of SPT N value for the case where the Silent Piler is used alone, or assisted by water jetting, or simultaneous inner augering and rotary cutting (Gyro Piler). Further details including the availability of various Press-in-Systems to cope with spatial constraints may be found in *Press-in Retaining Structures: A Handbook*, first edition 2016 published by the International Press-in Association (IPA, 2016).

The second aspect is to learn for what kind of infrastructure, by what reasons, in what circumstances, embedded wall and structures have been chosen as solution, and why they were constructed by the press-in piling. Project owners may be most interested in this aspect especially when they are dealing with existing infrastructure; to expand their capacity, to reinforce and recover the structural integrity of aged structures, to retrofit the structure to satisfy revised design requirements such as with a seismic retrofit, or to conduct emergency repairs of infrastructure damaged by natural and man-made disasters. The prime importance for the project owners in such projects is not the direct construction cost alone but also the indirect public cost during construction. For surface transportation, for example, project owners are concerned with indirect costs to commuters and the local economy due to disruption of services, detouring, and increased congestion during construction. Actually, the Silent Piler has often been selected to minimize the disruption or decreased performance of existing infrastructure.

The Japan Press-in Association compiled around 1,000 case records at its Japanese language site. Table 3.1 is reconstructed from the case records and summarizes the current applications of embedded walls together with illustrations. The following sections in this chapter provide typical examples of the most popular applications to date under the categories of urban development, surface transportation, and inland water system which may cover 70%–80% of current applications in Japan. The rest of the applications in Table 3.1 are not popular at the moment but are expected to increase in the near future. These emerging applications will be shown in Chapter 4.

Table 3.1 Typical applications of embedded walls

Typical applications and their characteristics	Illustration

Urban redevelopment and renovation:
 	
• Construction in narrow space
  • Improvement of open ditch and underground conduit in narrow working space.
• Maximum use of available land space
  • Temporary or permanent retaining wall with limited or practically no clearance against neighboring buildings.
• Headroom restrictions
  • Construction underneath viaduct or power lines
• Construction alongside busy traffic

Improvement of surface transportation: roads and railways:
 	
• Road widening/additional tracks
  • Retaining wall for depressed section and elevated section
• Grade separation to mitigate traffic congestion
  • Change one road to underpass another at surface crossing.
  • Necessary to minimize the impact on the traffic capacity during construction.
• Improvement of bridges: rehabilitation and retrofit
  • Cofferdam for rehabilitation of bridge piers
  • Steel tubular pile cofferdam foundation to increase bearing capacity of existing bridge foundation

(Continued)

Table 3.1 (Continued) Typical applications of embedded walls

Typical applications and their characteristics	Illustration
Inland water system: • Reinforcement of shore protection such as revetment or levees • Widening and deepening to increase the cross section of water channel	
Improvement of Maritime transportation: • Upgrading of navigation channel • Upgrading of breakwater • Improvement of quay wall and revetment	
Coastal protection: • Seismic retrofit of existing coastal levees • Restoration of coastal levees • Beach protection	
Seismic reinforcement of various infrastructures: • Tank foundation • Road and railway embankment • Underground utilities	
Bike commute infrastructure: • Underground bicycle stands and underground automated car park	

(Continued)

Table 3.1 (Continued) Typical applications of embedded walls

Typical applications and their characteristics	Illustration
Other interesting applications: • Preservation of historic structures • Protection of structures near the sinkhole • Sheet piling to control settlement due to adjacent construction	

3.2 Urban redevelopment and renovation

Noise and vibration, and any other adverse influence, both on nearby residents and structures are a major concern for the construction works, especially in urban areas. Construction constraints often encountered in urban areas are caused by a narrow working space between buildings or in a narrow street, construction adjacent to existing buildings or underground utilities, headroom restriction, underground obstacles, and construction along busy traffic. The press-in piling by the Silent Piler together with the Non-staging system is an indispensable technique in urban construction.

3.2.1 Construction in a narrow space

The Non-staging system comprising compact Silent Piler, clamping crane, and pile runner enables construction between houses (Figure 3.2) or in a narrow street where heavy equipment such as truck crane or crawler-mounted crane cannot enter.

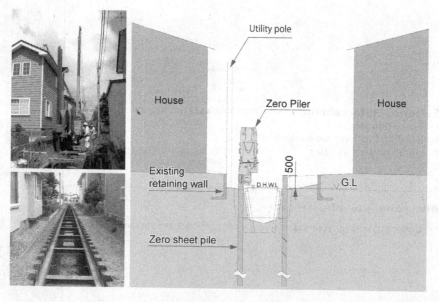

Figure 3.2 Construction in a narrow space between houses at Hokuto City, Hokkaido.

3.2.2 Adjacent construction maximizing the use of available land space

As the land cost is enormous in highly developed urban areas, the maximum use of available land space is often requested. Figure 3.3 compares the adjacent construction by the ordinary press-in method and a specially developed technique called the "Zero Clearance method". An ordinary Silent Piler is capable of installing a sheet pile wall alongside an adjacent building at a distance of 500–600 mm, which is determined by the width of the Silent Piler. Specially designed Zero sheet piles and a Zero Piler can reduce this distance to practically zero. Figure 3.4 shows an example of a sheet pile wall installation for a braced excavation.

3.2.3 Headroom restriction

Construction under a bridge, viaduct, or electric transmission line must be carried out within the given clearance. Height restrictions for equipment must also be strictly observed in construction around airports during the landing and take-off of aircraft.

Shown in Figure 3.5 is an example of the river wall renovation at Furukawa River, also known upstream as Shibuyagawa River, which is a small river flowing through a portion of down town Tokyo, collecting and draining precipitation from around 23 km². In order to avoid any adverse influence on the nearby structures and to minimize the reduction of cross section of the river during construction, a steel tubular

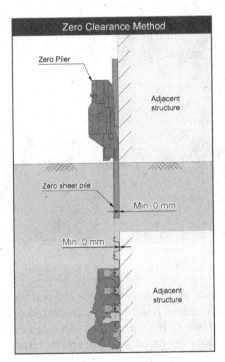
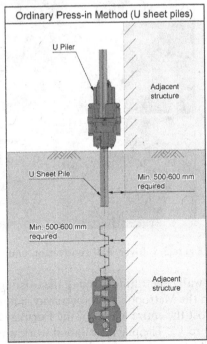

Figure 3.3 Comparison of ordinary press-in method and Zero Clearance method.

Figure 3.4 Braced excavation for a new building in Matsue City.

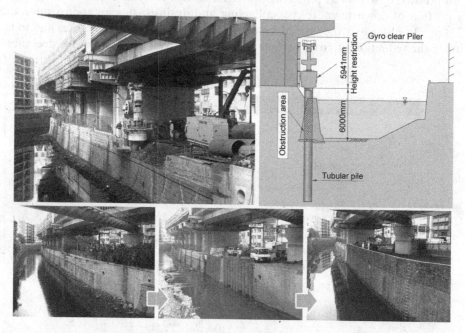

Figure 3.5 River wall renovation underneath viaduct.

pile wall penetrating through the existing river wall was selected for renovation. How-ever, the Metropolitan Expressway, a limited access road, passes adjacent to or above almost the entire length of the Furukawa River in the form of a viaduct as shown in Figure 3.5. The pile installation work was therefore faced with headroom restrictions.

The rotary press-in piling could penetrate through the existing reinforced concrete wall. The "Gyro-clear piler" which was specially designed for work under headroom restriction was selected to install the steel tubular piles (hereinafter called tubular piles) of 1,000 or 1,400 mm diameter.

3.2.4 Construction alongside busy traffic

In advancing construction alongside a busy traffic route, safety in construction is of the utmost importance because construction accidents such as crane contact with trains or vehicles, an overturned crane, or the dropping of loads present a risk of serious or fatal injury not only to workers but to the general public. It is also important to minimize the disruption to traffic during construction.

Figure 3.6 shows the pile installation work in the close vicinity of railway tracks. In these examples, the Non-staging system enabled pile installation within a limited space on the slopes. Stability of the Silent Piler and clamping crane during construction was guaranteed by the grip and self-walk mechanism. Accordingly, the construction was made possible without disturbing ordinary train services.

Figure 3.7 illustrates the cross section of the new Tokyo Metro Subway Line #12 at Okachimachi crossing. This particular construction site features railway tracks (JR East) crossing on a viaduct over a busy four-lane prefectural road (Kasuga Dori) at

a. Relocation of railway tracks

b. Reinforcement
of railway embankment

Figure 3.6 Construction in the close vicinity of railway tracks. (a) Relocation of railway tracks. (b) Reinforcement of railway embankment.

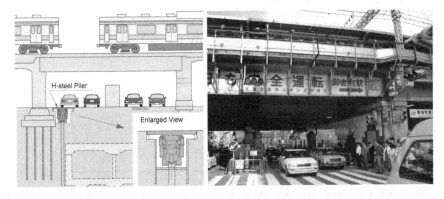

Figure 3.7 Braced excavation alongside busy road underneath railway viaduct.

grade level. A subway tunnel and new station were constructed by the cut and cover method as shown in the figure. The construction starts with the installation of an embedded retaining wall followed by the placement of a deck over the excavation trench to serve as a temporary road until the completion of the construction. One of the major requirements of the pile installation was obviously the limited headroom underneath the viaduct. Another requirement is that the pile installation should be carried out within a limited time at night because the traffic volume of Kasuga Dori does not permit even the partial closure of lanes during busy hours.

Figure 3.8 shows the construction sequence devised by taking the full advantage of compact Silent Piler. H-shaped steel sheet piles 25.5 m long were cut into five or six sections to enable press-in piling under the viaduct. At the embedded wall location a ditch 2.5 m deep and 2.15 m wide was prepared to accommodate the Silent Piler. During day time, the ditch was covered by a deck plate for traffic movement, leaving the piler in place, and at night, pile installation work was resumed immediately after removing the deck plate, which contributed to an efficient work schedule by avoiding any time for dis-assembling and reassembling of the piler. Figure 3.9 shows the pile installation work at night.

3.3 Surface transportation: Road and Railway

Congestion of the surface transport network is a common problem both for developing and developed countries. Adding lanes or railway tracks may ease congestion, but the scarcity of space for additional lanes/tracks is always the problem in urban areas.

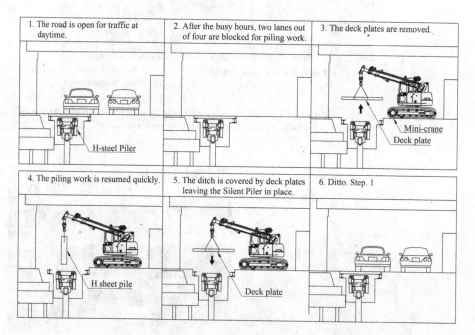

Figure 3.8 Construction sequence devised by taking the advantage of compact machine.

Figure 3.9 Pile installation at night.

Replacing the sloping sides of an existing transport corridor by constructing an embedded retaining wall to create a space for new lanes or tracks is common practice both for sections on embankment or in cutting in urban areas.

Congestion cannot be solved by road widening alone. Often a surface crossing or a deteriorated bridge becomes a bottleneck for smooth traffic flow and needs to be upgraded. Furthermore, it is highly recommended to minimize the total or partial closure of existing infrastructure which would invite increased congestion and/or detouring during construction.

3.3.1 *Widening of a road in cutting*

Hodogaya Bypass, connecting Yokohama and Tokyo, bypasses a part of National Route 16, a beltway of the Greater Tokyo Area. The bypass is a limited access road which carried more than 130,000 vehicles per day at the time of road widening in the Fiscal Year 2000. The purpose of road widening was to add one lane in each direction to divert the traffic at a proposed new interchange at Shin-Sakuragaoka. As shown in Figure 3.10, the site is in cutting and surrounded by residential buildings, so space for additional lanes could be acquired only by cutting the slopes on both sides. Cantilever-type embedded walls were selected to support the excavations. Requirements for construction were as follows:

- Low noise and low vibration.
- Walls should be embedded into stiff mud stone layer as shown in the boring log (Figure 3.10).
- Construction should be completed in a narrow working space to maintain the traffic volume.
- Pile installation on the slope required high stability of equipment during construction.
- Safety procedures in relation to roads and traffic.

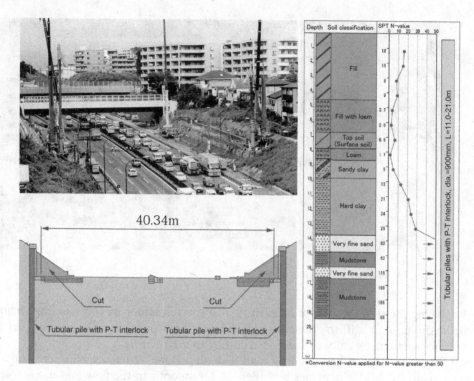

Figure 3.10 Road widening by excavation of slopes on both sides.

Four wall alternatives including Chicago Caisson Pile (φ 2,000 mm), cast-in-place concrete wall (φ 1,000 mm), pre-stressed concrete pile wall (500 mm x 500 mm), and interlocking steel tubular pile wall (φ 900 mm) were compared under these site-specific requirements in relation to their construction procedures. The Project owner, Japanese Ministry of Land Infrastructure and Transport, selected interlocking steel tubular pile wall and the press-in method based on the environmental and safety issues, construction speed, economy, and esthetics during and after construction.

The press-in method employed at this site comprised the Silent Piler and Non-staging system. Pile installation was conducted by static press-in piling assisted by the simultaneous inner augering for advancing piles down into mudstone with SPT N value varying from 60 to 167. The Non-staging system comprising the Silent Piler, clamping crane, and pile runner was employed to overcome the narrow working space on and near the slope (Figure 3.11). Tubular piles with interlock were brought onto the construction site in two pieces and welded together at the pile work yard. The pile runner carries the pile along the tops of the already installed pile row to the new pile location. The clamp crane picks up the pile, and the pile is pitched into the Silent Piler and pressed into the ground. The Silent Piler solved the noise and vibration problem. The Silent Piler and clamping crane gripping the piles guaranteed safety during construction. Figure 3.12 shows the completed road widening and the pile wall covered by the noise reduction fence.

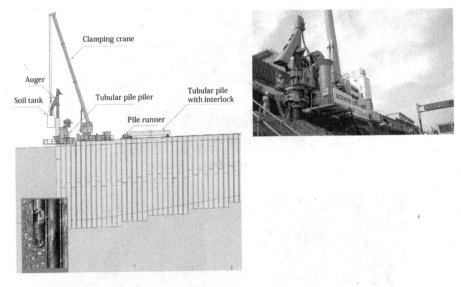

Figure 3.11 Pile installation by the Non-staging system.

Figure 3.12 Completed retaining wall covered by noise reduction fence.

3.3.2 Widening of elevated road on embankment

This is an example of a road widening project in Essex, UK. A cantilevered steel sheet pile wall was embedded in the slope to retain backfill which provides the space for an additional lane. Two Z-shaped steel sheet piles (hereinafter called Z sheet pile) AZ26 were interlocked in advance and installed by the Silent Piler down to 13–18 m to retain the slope and backfill. In this project, the sidewalk and one of the three active car lanes were used as a working space, which reduced the congestion during construction.

a. Pile installation in progress

b. Pile installation finished

c. Completed retaining wall

Figure 3.13 Press-in operation in progress and retaining wall construction sequence. (a) Pile installation in progress. (b) Pile installation finished. (c) Completed retaining wall.

Figure 3.13 shows the press-in operation in progress and the completed wall, and Figure 3.14 shows the construction plan and cross section.

3.3.3 Rail–Rail grade separation

The West Toronto Diamond, at-grade crossing of the Canadian National (CN) Railway and Canadian Pacific Railway (CPR), was one of the busiest rail intersections in Canada and a bottleneck for the increasing demands of freight and passenger trains. Shown in Figure 3.15 is the overall view of at-grade Rail–Rail crossing before the project. To improve service and safety by eliminating at-grade crossing, the CN tracks were depressed below grade and underpass the CPR tracks while the CPR tracks remained at grade at their current location. The depressed CN section is retained by interlocking steel tubular pile wall and the CPR tracks cross by a bridge which uses the steel tubular pile wall as its foundation. By well-considered project staging, track closure time was minimized to reduce the adverse impact on commuters. Further details of the project may be found elsewhere (e.g., Archibald et al., 2007).

The area for constructing the depressed section, which is shaded white in the figure, was in an area of mixed-use-zoning including industrial, commercial, and residential

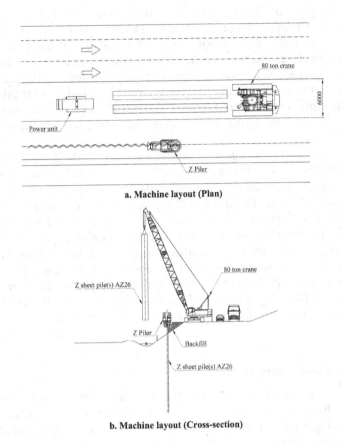

a. Machine layout (Plan)

b. Machine layout (Cross-section)

Figure 3.14 Machine layout plan and cross section. (a) Machine layout (plan). (b) Machine layout (cross section).

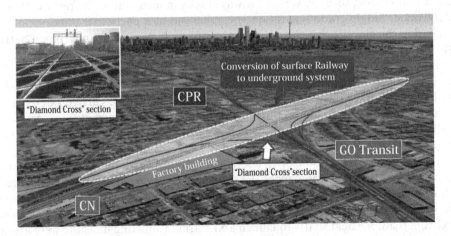

Figure 3.15 Overall view of the West Toronto Diamond Rail–Rail grade separation project.

buildings. The pile installation along the active railway demanded safe construction both for the construction workers and passing trains. The impact of noise and vibration due to piling work was a major concern for residents dwelling close to the construction site. In close proximity to sensitive structures, any adverse influence of piling had to be minimized, which was the major reason for adopting the press-in piling in this project. In the most critical area, the interlocking steel tubular pile wall was located at a distance less than 2 m from the footings of 100+-year-old factory buildings as shown in Figure 3.16. Along the factory building, the soil conditions were characterized as follows.

The upper soil layers, from ground surface to about 9 m deep, consisted of silty sand and clay with SPT N value in the range of 20–30. A very dense sand layer with SPT N value exceeding 50 appeared at a depth of about 9 m and was 2–3 m thick. The layer beneath the sand consisted of a Glacial Till (silt-clay) with blow counts that ranged from 30 to 90. The installation of 12–25 m long steel tubular pile with 914.1 mm diameter attached with P-T interlock was required. The press-in piling assisted by simultaneous inner augering was adopted to penetrate through the difficult soil conditions.

The piling work had to observe strict vibration limit. Extensive monitoring of the ground was conducted along the factory building by 12 inclinometers and 25 deep settlement rods. Direct monitoring of the factory building wall was also conducted by optical surveying. The press-in piling was indispensable for the completion of the project.

3.3.4 Rail–Road grade separation

According to the Japanese Ministry of Land, Infrastructure and Tourism, about 33,000 Rail–Road crossings exist in Japan in 2017. Such at-grade crossings invite increasing congestion, time loss, accidents, and CO_2 emissions, resulting in great economic losses. Grade separation of rail and road is being carried out either by continuous grade separation or single grade separation. Continuous grade separation can eliminate many crossings along a substantial length by elevating the railway tracks on viaduct or by lowering railway tracks in a cutting or by putting them underground. Single grade separation is carried out at a particular crossing, often by changing the elevation of the crossing road to an overpass or underpass relative to the railway tracks.

Figure 3.17 shows an example of continuous grade separation by elevating existing railway tracks. As the new railway track is constructed above or alongside the existing railway track, the construction of the new viaduct had to be carried out as close as possible to the existing railway track, while maintaining the current train services. So-called "Zero Piler" (see Figure 3.3) which can install the temporary retaining structure at a minimum clearance was employed.

In single grade separation, often a road passes underneath the railway track through a tunnel and the road approaches the tunnel in a depressed section, as was the case shown for the Rail–Rail crossing in the previous section. An embedded retaining wall is preferred in densely built circumstances to minimize any influence to traffic on the existing road. Vertical shafts to launch and extract tunneling machine were also constructed by the walk-on-pile-type press-in piling. Figure 3.18 illustrates a road underpass the railway track.

Figure 3.16 Embedded wall construction in the vicinity of old structure.

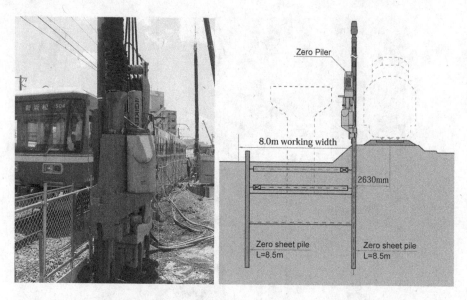

Figure 3.17 Piling works close to the active railway track for continuous grade separation (Ozawa Civil Engineering and Construction Co. Ltd).

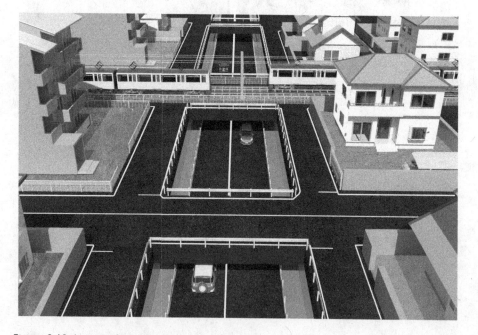

Figure 3.18 Image of single grade separation.

Figure 3.19 Image of Road–Road grade separation in densely built urban areas.

3.3.5 Road–Road grade separation

Common problems encountered in the Road–Road grade separation in a densely built urban environment are: a limited right-of-way, restriction of construction time and space, special care for minimizing any adverse influence on the current traffic volume during construction, and noise and vibration. Grade separation can be achieved by depressing the major road with the larger traffic volume underneath minor roads with less traffic, thereby permitting through traffic on the major road to travel without traffic signals as exemplified in Figure 3.19. By employing embedded walls for the retaining walls of the depressed road and for the foundation of the bridge structures of the at-grade road, the construction sequence is simplified and space required for construction is minimized by the use of the walk-on-pile-type press-in piling.

3.3.6 Bridges

Old bridges, built decades ago, were designed based on the then available design code and for the traffic load predicted at the time of construction. Upgrading of aged bridges or restoration of damaged bridges is necessary to maintain or recover their function. Seismic retrofitting of old bridges is being carried out to meet the current design code revised following the experience of huge earthquakes in the past decades. Depending on the condition of a bridge, necessary measures are taken for the superstructure, bearings, bearing seats, abutments, or substructures such as piers, piles, or footings. When traffic volume exceeds the capacity of an existing bridge, construction of a new bridge near the existing bridge is undertaken in addition to the upgrading of existing bridge.

To deal with bridge piers, footings, or other bridge foundations, it is often necessary to create dry working condition by constructing a temporary cofferdam using steel sheet piles or tubular piles with interlock under the existing superstructure. In addition to headroom restrictions, there are further requirements depending on the expected function of the river:

- Complete construction within the dry season
- Minimize the reduction of river cross section
- Minimize the adverse influence on navigation
- Minimize noise and vibration

The walk-on-pile-type press-in piling is superior to other piling techniques in satisfying these requirements by the use of compact equipment together with the Non-staging system enabling pile installation and extraction after construction, even within extremely limited headroom.

Steel sheet piles or tubular piles are sometimes left in place to increase the horizontal and vertical load carrying capacity of the existing foundation system, by connecting the pile wall to the existing foundation.

3.3.6.1 Retrofitting of Kuramae Bridge

In 1923, the Great Kanto Earthquake of magnitude 7.9 devastated Tokyo Metropolitan area resulting in more than 100,000 casualties. Most of them died as a result of the fire that swept across the city and the lack of any evacuation route. Figure 3.20 shows

a. Present view of Kuramae Bridge

b. Kuramae bridge just after construction in 1927

Figure 3.20 Kuramae Bridge over Sumida River. (a) Present view of Kuramae Bridge. (b) Kuramae bridge just after construction in 1927 (Japan Society of Civil Engineers, JSCE Library).

the Kuramae Bridge over Sumida River. The bridge was built in 1927 as a part of the restoration and reconstruction from the mega earthquake.

The three-span steel arch bridge is supported by three reinforced concrete wall piers founded on timber piles and an abutment. The bridge was designated as an important historical and cultural property by the Tokyo Metropolitan Government but still plays an important role in surface transportation, carrying more than 20,000 vehicles per day.

Retrofitting of the bridge foundation undertaken in 2003 is described below. Figure 3.21 shows the cross section and the plan of retrofitting work on the P1 pier, and Figure 3.22 shows the typical soil profile. The major construction sequence is: pile installation to form a closed perimeter around the existing footing, excavation of the river bed within the pile wall, drilling and bonding 1,080 steel bars into the existing footing, connecting the pile wall and existing footing by adding a new concrete slab, and finally cutting off the pile walls at the top level of the footing. As described above,

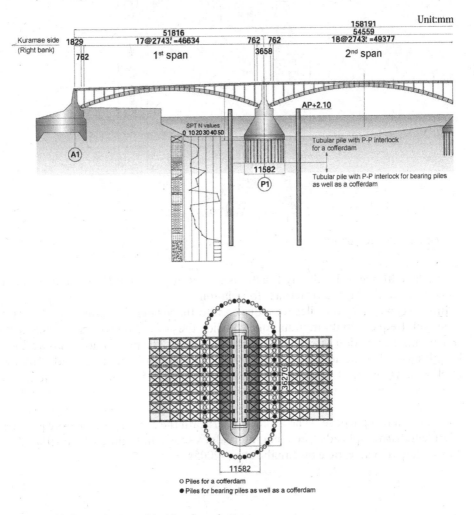

Figure 3.21 Retrofitting of bridge foundation.

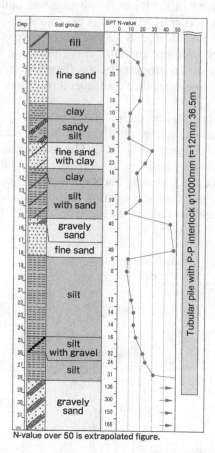

N-value over 50 is extrapolated figure.

Figure 3.22 Soil profile.

the peripheral pile wall initially functions as a cofferdam and later becomes the permanent steel tubular pile cofferdam foundation.

Initial four temporary piles were driven into the ground by means of a hydraulic-type high-frequency vibration hammer outside the cofferdam so that the Silent Piler can obtain the reaction force for the production piles. The production piles were tubular piles with P-P interlock, the size of which is 1,000 mm in diameter and 12 mm wall thickness. Among the 84 production piles, 28 piles are 36.5 m long reaching to the sand and gravel layer to obtain improved bearing capacity (shown by solid circles in Figure 3.21) and 56 piles are 14.5 m long to form a continuous embedded wall primarily for the water shutoff purposes. Those piles underneath the existing bridge were pressed-in short lengths and spliced at frequent intervals as shown in Figure 3.23. Further details may be found in an article by Tanabe et al. (2005).

The following labels appear on the figure diagram: 2200mm, 2370mm Height restriction, A.P +2.10, A.P+2.00, Cut-off level ▽A.P-4.10, ▽A.P-6.40

Figure 3.23 Pile installation underneath the bridge.

3.3.6.2 Improvement of bridge capacity (combined project of old bridge rehabilitation and new bridge construction)

Figure 3.24 shows the location of three bridges, Kanchpur Bridge, Meghna Bridge, and Gumti Bridge, on Bangladesh's National Highway No. 1 (NH-1) where a project to rehabilitate existing bridges and construct new bridges in parallel with existing ones is ongoing. The existing bridges were constructed in 1977, 1991, and 1995, respectively. NH-1 connects Dhaka, capital and the largest city of Bangladesh, and Chittagong,

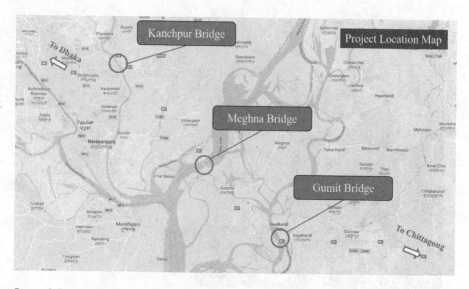

Figure 3.24 Location of three bridges on the National Highway No.1, Bangladesh. (Map: https://www.google.co.jp/maps/.)

the port city and the second largest city. There are three combined reasons behind this project: mitigation of traffic congestion, rehabilitation of bridge structures, and seismic retrofitting.

Along with the economic growth of the country, the amount of freight traffic and passenger transport has drastically increased and exceeded the capacity of NH-1. To mitigate the traffic congestion, the Government of Bangladesh has been widening all sections of NH-1 into four lanes from the existing two lanes except for bridges. As the bridges are becoming bottlenecks for the NH-1, the government decided to add four lanes at the abovementioned river crossings by constructing new bridges. The existing bridges have been deteriorating since several years and need urgent rehabilitations. Furthermore, as the existing bridges were designed based on the outdated standard, they need seismic retrofitting to satisfy the current seismic codes. Although the rehabilitation and retrofitting include the works on superstructure and pier structure, the current article focuses upon the foundation alone.

The optimum route of new bridges was determined as immediately next to the existing bridges, either on downstream side (2nd Kanchpur Bridge and 2nd Gumit Bridge) or upstream side (2nd Meghna Bridge), by considering various viewpoints including cost, technical issues, and environmental and social issues.

Permanent steel tubular pile cofferdam foundation and cast in situ concrete piles were chosen as candidates for the foundation of the new bridges and the retrofitting of the existing bridges. Figure 3.25 shows the conceptual image of two types of foundation.

The steel tubular pile cofferdam foundation has been adopted effectively for retrofitting old bridge foundations as shown in the previous section. The foundation system is also applicable for new bridge foundation. As the project involves both retrofitting of

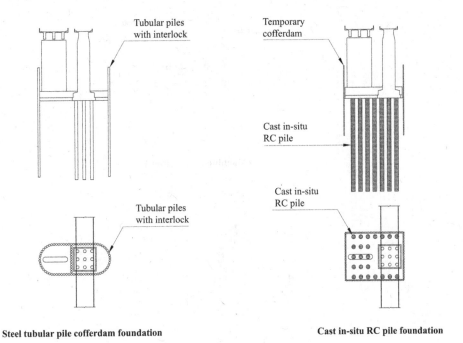

Figure 3.25 Image of steel tubular pile cofferdam foundation and cast *in situ* RC pile foundation.

existing bridges' substructures and constructing foundations for new bridges immediately next to the existing bridges, interlocking tubular pile wall enclosing both existing and new foundations was proposed as shown on the left in Figure 3.25. The pile wall functions as the cofferdam to create dry working condition in the early phase and becomes an essential part of foundation after it is joined to the concrete slab.

In the case of cast-in-place concrete pile foundation shown on the right of Figure 3.25, a number of additional piles and the expansion and strengthening of concrete slab will be provided on the existing bridge foundation to meet the rehabilitation and retrofitting purposes. Immediate next to the existing foundation, new piles will be installed for the foundation for a new bridge. Furthermore, the foundation system requires a peripheral cofferdam just for the temporary purpose.

The final selection of the foundation type is determined for each pier based on the soil conditions, bridge type, and geometry including design water depth. In the current project, the steel tubular pile cofferdam foundation is the preferable option for most of the piers as long as the water depth including the scouring depth is deep according to the annex of the referenced material (Government of the People's Republic of Bangladesh, Ministry of Road and Transport and Bridges, 2014). Where the water depth becomes shallower, the piled foundation becomes competitive or advantageous.

The project started in December 2015 and is expected to complete after 7 years and 8 months. At the time of preparing this book, the installation of the steel tubular pile cofferdam foundations has been completed.

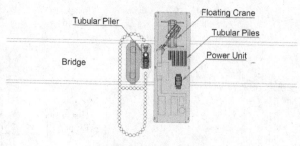

a. Machine layout (Plan)

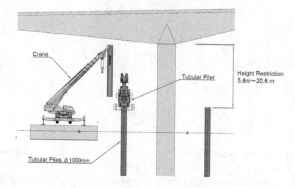

b. Machine layout (Cross-section)

Figure 3.26 Pile installation at deep water underneath the existing bridge. (a) Machine layout (plan). (b) Machine layout (cross section).

Figure 3.27 Pile installation underneath the existing bridge.

Figure 3.26 shows the layout of equipment for pile installation on the water under the headroom restriction. Figure 3.27 shows the pile installation work underneath the existing Gumti Bridge.

3.4 Inland water

Inland water systems such as rivers, canals, and channels have various functions and each is managed accordingly. The focus in this section is inland water in locations where construction space is limited due to other land use.

One of the important functions of inland water is flood control by retaining and discharging precipitation. Flooding and inundation caused by increased frequency of heavy rainfalls, melt water, and intensified typhoons/hurricanes/cyclones, together with increasing sea levels, are common concerns in many areas around the world. Expanding the cross section of rivers and canals is one of the solutions among others to reduce the risk to human lives and properties in urban areas. Inland water, especially big rivers or canals, provides an economical means of transport which requires appropriate width and depth for the navigation of vessels. In the examples shown earlier for bridges, the rivers flowing underneath the bridge are utilized for navigation. Inland water is also used as a source of tap water. In any case, it is important to secure the cross section of the river necessary for its respective functions.

Excavation to increase the cross section of a waterway often requires improvement of deteriorated existing levees and revetments. Space for a construction yard on the protected side is often limited. However, the use of space in the river is generally restricted to the dry season or prohibited because it would result in the reduction of the existing function of the waterway, whether for discharge, water conveyance, or water supply. An embedded wall created by steel sheet piles or tubular piles penetrating through the existing protections are preferred solution, and hence, river renovation by an embedded wall is one of the most popular fields of application for the walk-on-pile-type press-in piling and about 400 cases have been reported, mostly in Japan but also from North America and Europe.

3.4.1 River renovation work at Zenpukuji River

Zenpukuji River runs through a low-lying residential area of Tokyo and discharges rain water to Tokyo Bay via Kandagawa River downstream. The residents along the river have often suffered from flooding during intense rain fall or typhoons. As the residential buildings on both sides are very close to the existing river revetment, noise and vibration are prohibited during construction. Furthermore, it is almost impossible to secure a working space for the renovation. The selected solution is to install the cantilever earth retaining wall by tubular piles just behind the existing revetment as shown in Figure 3.28.

Tubular piles with 1 m diameter, 16–16.5 m long were installed down into sand and gravel layers to maintain stability (Figure 3.29). To cope with the narrow working space, the material stock yard was secured over the river at the starting point of the construction. The combination of two units of Silent Piler assisted by the rotary cutting and Non-staging systems was employed to complete the renovation work as shown in Figures 3.30 and 3.31.

3.4.2 Renovation and restoration of drainage channels in North America

The use of the press-in piling for the improvement of existing drainage channels has been accepted gradually in North America. The reasons of selecting the walk-on-pile-type

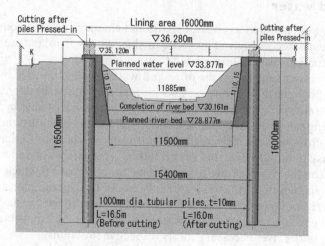

Figure 3.28 Cross-section of Zenpukuji River.

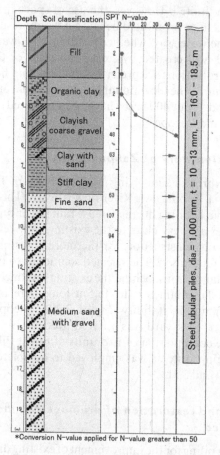

*Conversion N-value applied for N-value greater than 50

Figure 3.29 Typical soil profile of Zenpukuji River.

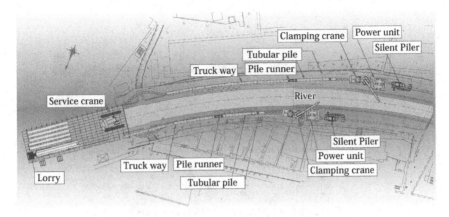

Figure 3.30 Construction plan of embedded wall at Zenpukuji River.

Figure 3.31 Pile installation in progress.

press-in piling are low noise and vibration, the capability to penetrate into stiff soil layers, and mobility in narrow construction areas.

Figure 3.32 shows an improvement to a water drainage channel in Orange County, California. Z sheet piles, PZC26, 45 feet long were driven into the existing levee to create a double sheet pile flood wall. Construction besides a residential area demanded noise- and vibration-free pile installation. Two to three units of Silent Piler were mobilized simultaneously to minimize construction time.

3.4.3 Renovation of aqueduct for water supply

Musashi Canal, 14.5 km long, connects Tone River and Arakawa River. The Canal constructed in 1967 is one of the facilities operated and maintained by the Japan Water Agency to supply one-third of the demand for drinking quality water to Metropolitan Tokyo. To cope with the degradation of the Canal due to aging and ground subsidence, the renovation of the Canal was undertaken. The renovation work involved the modification of the current single Canal to create two parallel water channels, which was necessary to mitigate the influence of disruption of service during construction

a. Before construction   b. After construction

c. Under construction

Figure 3.32 East Garden Grove – Wintersburg Channel, Orange County, CA, USA.
(a) Before construction. (b) After construction. (c) Under construction.

by allowing one channel to continue water supply while the other channel was under construction. This dual channel system is also considered effective for the future maintenance of the Canal.

The renovation work therefore required a temporary water cutoff on the center line of the Canal by installing an embedded wall. One of the critical points for the renovation was at the crossing of the Canal and the railway tracks, where the clearance beneath the railway bridge and water surface was merely 1.5 m. Figure 3.33 shows the pile installation under the railway bridge, and Figure 3.34 shows the successful water cutoff on one side allowing the other side to continue the water supply.

When the headroom is insufficient to allow threading of pile in full length, a pile is pressed-in by successive installation of short piles. A short pile is pitched from the vertical direction to connect with the adjacent pile at the interlock and then welded to the previously installed short pile to establish vertical continuity. In the present case of exceptionally tight headroom, the above procedure is no more practicable.

A special interlocking device was developed to permit the connection from the lateral direction as shown in Figure 3.35. Also a vertical mechanical joint (Figure 3.36) was developed to establish vertical continuity, avoiding the difficult welding work in the flowing water. A new Silent Piler was developed that was compatible with the new sheet pile geometry as shown in Figure 3.37. The process of pile installation by new method is schematically shown in Figure 3.38. Arai et al. (2018) describe further details of the new method.

The renovation work was successful and the Japan Society of Civil Engineers awarded the project as an Outstanding Civil Engineering Achievement of 2017.

Figure 3.33 Pile installation into Musashi Canal under extreme headroom restrictions.

Figure 3.34 Dry work on one side and water supply on the other side of the Musashi Canal.

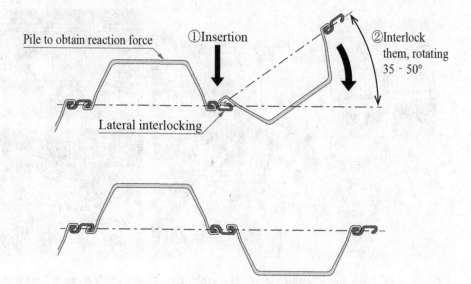

Figure 3.35 Special interlocking device for horizontal feed joint.

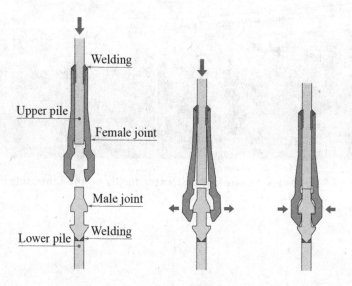

Figure 3.36 A mechanical joint to establish vertical continuity.

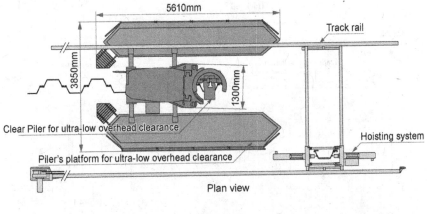

Plan view

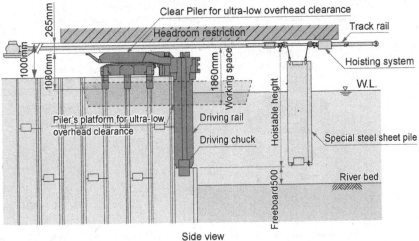

Side view

Figure 3.37 Specially designed Silent Piler & sheet pile to cope with extremely limited headroom.

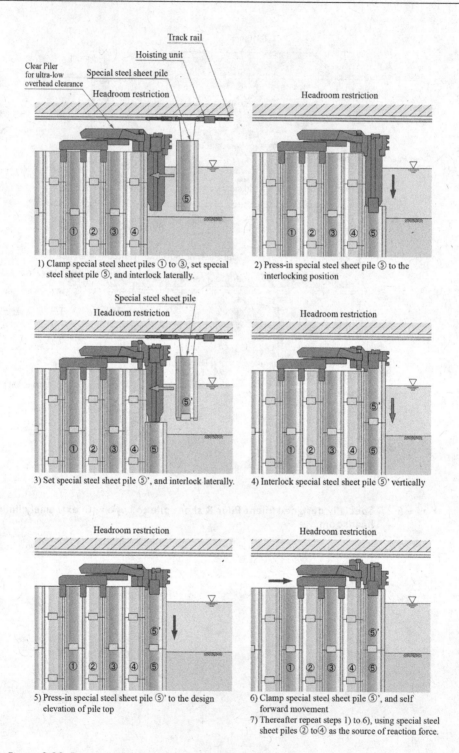

1) Clamp special steel sheet piles ① to ③, set special steel sheet pile ⑤, and interlock laterally.

2) Press-in special steel sheet pile ⑤ to the interlocking position

3) Set special steel sheet pile ⑤', and interlock laterally.

4) Interlock special steel sheet pile ⑤' vertically

5) Press-in special steel sheet pile ⑤' to the design elevation of pile top

6) Clamp special steel sheet pile ⑤', and self forward movement

7) Thereafter repeat steps 1) to 6), using special steel sheet piles ② to④ as the source of reaction force.

Figure 3.38 Process of press-in operation by new method.

References

Arai, M., Sugio, S., Murayama, K., Kunitomi, H. and Kimura, Y. (2018) Case study of underwater Press-in method of steel sheet pile under restricted headroom beneath a railroad bridge, *Proceeding of 1st International Conference on Press-in Engineering*, pp. 533–540.

Archibald, B., Anderson, V., Crabb, J. and Werner, J. (2007) West Toronto Diamond, *Paper Prepared for 2007 Annual Conference of the Transportation Association of Canada*, Saskatoon, Saskatchewan, 12 p.

Government of the People's Republic of Bangladesh, Ministry of Road and Transport and Bridges (2014) Implementation of resettlement action plan under the Knchpur, Meghna and Gumti 2nd bridges construction and existing bridges rehabilitation project (KMG Project) Revised resettlement action plan, http://www.rhd.gov.bd/RHDNews/Docs/RRAP_KMGP_June_2015.pdf.

IPA (2016) *Press-in Retaining Structures: A Handbook*, first edition, International Press-in Association (Second edition will appear at the end of 2020).

Tanabe, K., Itami, N. and Nishida, Y. (2005) Retrofitting of Famous Kuramae Bridge, *Doboku Sekou*, Vol. 46, Issue 2, pp. 94–99 (in Japanese).

Chapter 4

Emerging applications

4.1 Introduction

Following the previous chapter, this chapter provides the applications of embedded walls and structures, which are not popular at the time of writing this article but expected to increase in a near future. The categories of application covered in this chapter are maritime transport, coastal protection, beach protection, seismic reinforcement, bike commute infrastructure, and other interesting applications.

4.2 Maritime transport infrastructure

A port is a key component of maritime transport which supports daily lives of people. Most of crude oil and petroleum products, coal, grains, and raw materials for industries, foods, and variety of products pass through ports. In the events of large-scale natural disasters, maritime transport is indispensable for transporting emergency supplies in large quantity to the affected areas.

Expanding world economy and increasing domestic/international trade have demanded the continuous maintenance and upgrading of maritime transport infrastructure. The higher efficiency of maritime transport can be achieved by improving connection with rail and highways on land, introducing larger vessels, renovating port facilities such as cargo handling equipment according to changing cargo types, and improving port, breakwaters, and navigation channels on water side.

Larger vessels in terms of beam (breadth), length, draft, and height oblige port to acquire increasing length of the berth, deepening of water in front of the berth, deepening and widening of navigation channel, mooring space, and turn around basin. The recent expansion of the Panama Canal will increase number of larger vessels and necessitate improvement of maritime transport infrastructure.

4.2.1 Upgrading of navigation channel: an example at Miike Port

The Miike Port opened in 1908 for loading coal from Miike coal mine and contributed to Japan's industrial development. The port is a constituent of World Heritage "Sites of Japan's Meiji Industrial Revolution: Iron and Steel, Shipbuilding and Coal Mining". Even after the closure of coal mine in 1997, the port has continued to play an important role in the local economy.

The Miike Port locates in a large bay called Ariake-Kai in the Kyushu Island, which is relatively shallow near the shoreline with the large tidal range between 5 and 6 m. The port is approached from deeper water through the dredged navigation channel. Initially the navigation channel was protected by the sediment control groins on both sides to minimize siltation. Currently the navigation channel locates in between the sediment control groin in the north and the sea wall of a reclaimed land in the south as shown in Figure 4.1.

Since 1999, the upgrading of the international terminal has been in progress to accommodate larger container vessels and bulk carriers. The project includes improvement of land side facilities, deepening of berthing area and basin inside the port, as well as the deepening and widening of approach navigation channel. Upgrading of navigation channel was reported by Hiiragi (2006) and Kimura (2012).

The width and depth of approach navigation channel before upgrade were 50 m and −7.3 m, respectively, as shown by the broken line in Figure 4.2. The width and depth of 72 and −10 m were planned to accommodate 12,000 DWT (Dead Weight Tonnage) class vessels irrespective of tide at any time. It was necessary to construct embedded cantilevered walls to maintain the stability of sediment control groin and sea wall at both sides of the channel before dredging down to −10 m. Three thousand sixty-six steel tubular piles (hereinafter called tubular piles) with P-T interlock having 900 mm diameter and 10–11 mm thickness were to be installed from sea bottom at around 0 m down to −19.5 to −20.5 m all along the sediment control groin and sea wall as shown in Figure 4.3.

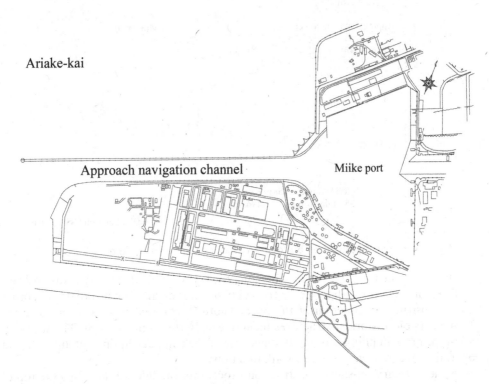

Ariake-kai

Approach navigation channel

Miike port

Figure 4.1 Miike Port and approach navigation channel.

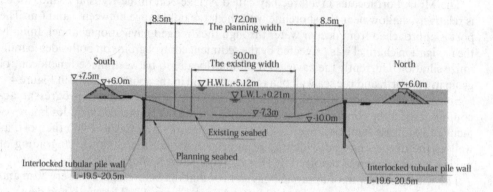

Figure 4.2 Deepening and widening plan of the navigation channel (Kimura, 2012).

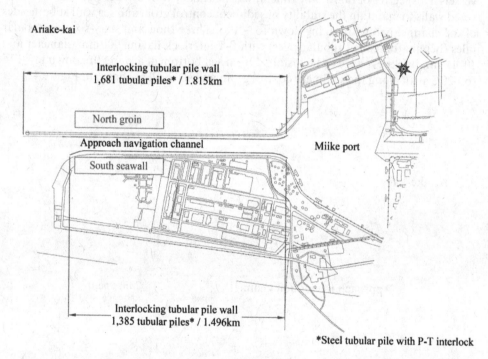

*Steel tubular pile with P-T interlock

Figure 4.3 Plan layout of embedded walls along navigation channel.

An important requirement for the construction of embedded walls was to leave sufficient navigable width to secure the safety of ships during pile installation in order not to obstruct the function of the port. Another requirement was to preserve the 100+ years old historical structures including sediment control groin. The walk-on-pile-type press-in piling was adopted because of the compact equipment and safety in operation due to its grip and self-walk functions.

In the ordinary on-shore construction procedure, the Silent Piler grips the top of previously installed piles to gain reaction force for penetrating a new pile and walk on the top of previously installed piles. If the ordinary procedure were followed, pile top level must be above the high water during pile installation and each pile has to be cut under water at the design elevation after the Silent Piler moved away. To achieve efficiency and economy, specially designed driving attachments were introduced as shown in Figure 4.4. The driving attachment is a 7.3 m long temporary extension tube having the same diameter with the tubular piles. The bottom 80 cm of the attachment equipped with hydraulic clamping system is inserted into the previously installed tubular pile to transmit reaction force at the bottom and allow Silent Piler to operate on the top of driving attachments above the high water.

Two flat bottom barges accompanied the Silent Piler during construction; one for mounting power unit, crane, and auxiliary equipment and the other for mounting tubular piles. Figure 4.5 shows the pile installation in progress while leaving space for ship movement.

When the construction of embedded wall was approaching to the tip of sediment control groin in the final stage of construction, a thick layer of rubbles (1 m diameter at maximum) was found near the light house. As the removal of rubble layers might endanger the stability of light house, 55 tubular piles equipped with interlock were replaced by the tubular pile of the same size with the cutting teeth at the bottom ends. These piles were successfully penetrated through the rubble layer by the rotary press-in piling (Gyro Piler) as shown in Figure 4.6.

4.2.2 Upgrading of existing breakwater: a new proposal of employing embedded walls

The harbor area is almost always protected by breakwaters against incoming waves to keep the calmness of internal water and to maintain the required basin depth. The breakwater also has a function to minimize the risk on human lives and assets in

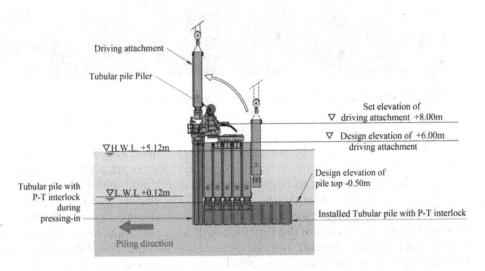

Figure 4.4 Use of driving attachments for efficient pile installation.

Figure 4.5 Pile installation in progress.

Figure 4.6 Installation of tubular piles into the rubble layer by Gyro Piler.

inland region behind a port by reducing the energy of incoming waves such as tsunamis. In most cases in Japan, composite-type breakwaters are preferred which are usually composed of upright sections (concrete caissons) and rubble mounds. Upright sections are seated on rubble mounds with small embedment and resist horizontal force. This is a typical gravity-type structure. The 2011 Great East Japan Earthquake damaged many breakwaters of this type especially under associated tsunami overflow.

Based on the lessons learned from the tsunami disaster, the revised design standard in Japan for shore protection facilities including breakwaters requires to upgrade existing facilities more resilient against next earthquake–tsunami event. Shore protection facilities are those facilities that protect human lives and assets from disasters due to tsunami, storm surge and waves, and movement of seabed which include breakwaters, jetties, revetments, parapets, seawalls, pumping facilities among others. The concept of resilience for the shore protection facilities is described in such a way that the risk of total collapse of facilities (wash away) should be minimized and the time to failure should be delayed as long as possible even when the tsunami height would exceed the design tsunami. The resiliency is expected to provide people additional time for evacuation, minimize the damage due to second or third tsunami attack, and enable quick restoration.

A simple reinforcement of breakwater is widening rubble mound toward the protected side and placing stabilizing berm against concrete caisson as shown in Figure 4.7a. Although the passive earth pressure by stabilizing berm can effectively improve the safety of breakwater against overturning and sliding, it may lead to total failure under the action of extraordinary horizontal load. A drawback for the port users of this reinforcement is that the reinforcement work interferes the ship movement during construction, and the additional stabilizing berm would reduce the port function permanently after the construction as shown in Figure 4.7b.

To overcome these disadvantages of reinforcement with additional rubble mound, the new type of reinforcement by installing steel tubular pile walls in the existing rubble mound and filling rubbles between the caisson and the pile wall was proposed and developed by Nippon Steel & Sumitomo Metal Corporation (currently known as NIPPON STEEL CORPORATION), jointly with the Port and Airport Research Institute, Tokyo University of Science, and the Coastal Development Institute of Technology. The image of reinforcement of composite breakwater by embedded wall is shown in Figure 4.8.

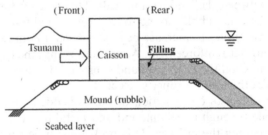

a. Reinforcement of breakwater by stabilizing berm

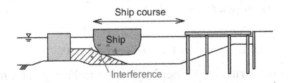

b. Reduction of ship maneuverable space due to additional berm

Figure 4.7 Reinforcement of breakwater by additional stabilizing berm (Moriyasu et al., 2016). (a) Reinforcement of breakwater by stabilizing berm. (b) Reduction of ship maneuverable space due to additional berm.

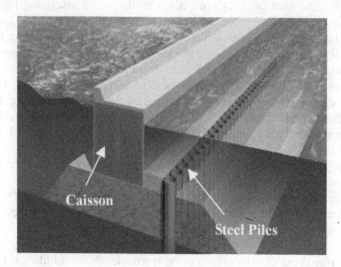

Figure 4.8 Image of breakwater reinforcement by steel tubular pile wall (Moriyasu et al., 2016).

Large-scale model tests in dry condition (Kikuchi et al., 2015) and hydraulic model tests to simulate tsunami overflow (Arikawa et al., 2015) were conducted and summarized by Moriyasu et al. (2016). The performance of the breakwater reinforced by embedded wall is confirmed as follows. The resistance to the horizontal wave force is initially carried by the base resistance of the caisson, but with increasing displacement of the caisson the horizontal force is transmitted to the embedded wall through the rubble infill between the caisson and wall. This mechanism changes the failure mode from sliding on the rubble mound to the deep-seated failure mode and the horizontal resistance continues to increase beyond the maximum resistance expected on the breakwater without reinforcement. Also found is that the caisson will not exhibit total collapse even when the rubble mound is eroded due to tsunami overflow if the protruded length of the wall is properly selected.

The difficult problem in construction of the embedded wall is the need of penetrating tubular pile through rubble mound and installing it to the required embedment without damaging the pile toe. This problem can be solved by employing the rotary press-in piling (Gyro Piler). The performance of the Gyro Piler was confirmed at Miike Port where rubble layers of about 7 m thickness were successfully penetrated as shown earlier in Figure 4.6.

4.2.3 Improvement of quay walls and revetments

A quay or wharf is a structure for berthing ships and cargo handling. Sufficient land space should be acquired behind the quay for locating functions necessary for maritime transport including temporary storage yard, warehouses, and access to road and/or railway. Shore line protection of these land space is called revetment or sea wall. Quay walls and revetments have been improved by various purposes such as deepening of harbor to accommodate larger ships, reinforcing to increase earthquake resistance, and restoring from aging or damages. The walk-on-pile-type press-in pil-ing is gradually becoming popular in such improvement works.

Construction of a new embedded retaining wall in front of or behind the existing quay wall has been one of the preferred improvement techniques. The new retaining wall may be cantilevered or anchored depending on site-specific conditions including the retaining height, embedment, ground condition, and section modulus of selected wall material.

Figure 4.9 shows sheet pile installation in front of a quay wall for small vessels at La Grande Motte, France. U-shaped steel sheet piles (hereinafter called U sheet piles) PU12 around 8 m long were installed by the Silent Piler to prevent the adverse influence of piling work on nearby structures.

Figures 4.10 and 4.11 show the renovation of −10 m quay wall at Yokohama Port. Tubular piles with P-T interlock, with 1,200 mm diameter and 15 mm thickness, were installed in front of the existing aged quay wall. The need to penetrate through rubbles on the sea bottom and the need to embed large diameter tubular piles 5–6 m into mud

Figure 4.9 Improvement of quay wall at La Grande Motte, France.

Figure 4.10 Renovation of −10 m quay wall at Kanazawa District, Port of Yokohama.

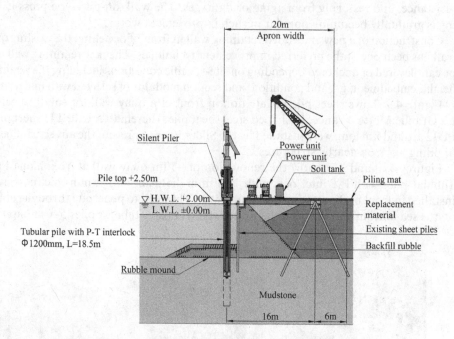

Figure 4.11 New steel tubular pile wall and aged existing quay structure.

rock layer, the press-in piling assisted by simultaneous inner augering was selected for this project. The same photograph shows a vessel is berthing at the adjacent berth which shows one of the benefits of renovation by embedded wall structure. While one berth is under construction, adjacent berth can continue providing services of berthing and cargo handling.

4.3 Coastal protection in earthquake-prone areas

The coastline of Japan has experienced earthquake and earthquake-induced tsunami disasters. Especially the 2011 Great East Japan Earthquake and tsunami disasters revealed the vulnerability of gravity-type coastal protections. Based on these experiences, design concept for the protection has changed, ensuring serviceability for earthquakes with high probability of occurrence during design life and reparability for earthquakes with low probability. Reinforcement and upgrading of existing protection by a resilient structure such as embedded double sheet pile walls or embedded tubular pile walls are becoming popular.

4.3.1 Reinforcement of coastal levee to prepare for future earthquakes and tsunamis

For the event of a huge earthquake at plate boundary off the coast, a series of events and their consequences must be examined carefully. In the 1946 Nankai Earthquake, due to the displacement and deformation of the earth's crust, Kochi City settled 1–2 m and strong earthquake motion attacked the city almost simultaneously and then increased water level arrived at the damaged shore protections. Most of the Kochi city was inundated for about a month period and 200,000 people were affected.

Along the coastline of Kochi Prefecture, existing coastal protections are being reinforced to cope with anticipated huge earthquake and tsunamis in near future. An example shown in Figure 4.12 is a reinforcement of coastal levee at Nino coast, Kochi Prefecture. The existing protection is vulnerable to earthquake-induced liquefaction and may lose its function before the arrival of tsunami. The selected solution here is reinforcing the existing sea wall by double steel sheet pile walls (Ishihara et al., 2020).

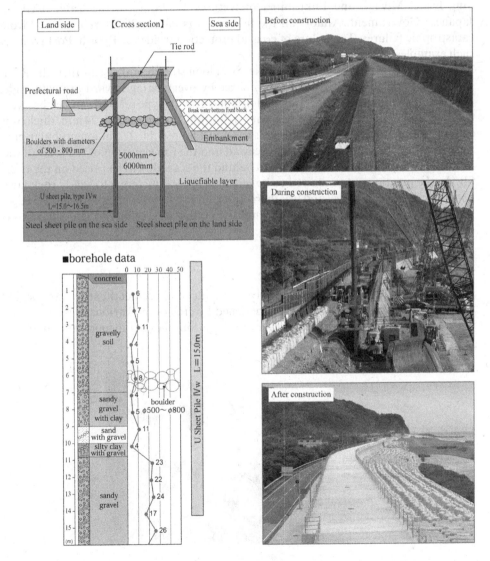

Figure 4.12 Double sheet pile wall to reinforce coastal levee.

As shown in the figure, U sheet piles are installed through liquefiable layer down to non-liquefiable layer, so that the protection maintains its design height in the event of huge earthquake and tsunami overflow. The press-in piling assisted by simultaneous inner augering was selected to reduce the adverse influence of vibration and to penetrate existing boulder with diameter ranging from 500 to 800 mm underneath the current sea wall.

4.3.2 Restoration of coastal defense damaged by tsunamis

The length of the totally or partially collapsed coastal defense sections due to 2011 Great East Japan Earthquake Tsunami reached 190 km out of 300 km coast line spanning Iwate, Miyagi, and Fukushima prefectures according to the Cabinet Office, Japanese Government. Almost all the fishery ports (about 260 ports) suffered from catastrophic failures. The failure of coastal protection inside the Ryoishi Port is one of such examples.

In the 2011 Great Tohoku Earthquake, 80 m long stretch of the seawall at the Ryoishi Port was failed under tsunami wave force by overturning toward protected side. Figure 4.13 shows the failed section and original gravity-type seawall can be observed. Cantilevered wall by tubular piles with 1,500 mm diameter and 24 mm thick was selected for upgrading the seawall. As shown in Figure 4.14, 27.5 m long tubular piles were installed in front of the original seawall and the top elevation of the seawall was increased from +9.3 to +12 m. The road behind the seawall will be raised on the backfill. As the pile tips were required to be installed several meters into sandy rock layer penetrating through gravel and weathered rock layers, tubular piles with cutting edges were used and pressed in by the rotary press-in piling (Gyro Piler). Figure 4.15 shows the pile installation work in progress.

4.4 Beach protection

A wide beach can dissipate energy of waves and tides and protect the structures behind the beach. Such beaches have been threatened by erosion by various reasons including reduced supply of sand from rivers or from longshore transport, the increased

Figure 4.13 The gravity-type seawall failed by overturning at the Ryoishi Port.

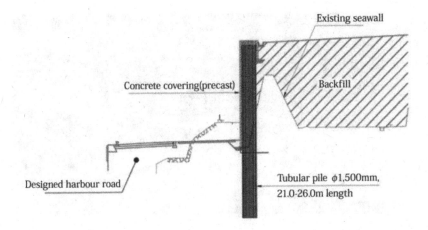

Existing seawall

Concrete covering(precast)

Backfill

Designed harbour road

Tubular pile φ1,500mm,
21.0-26.0m length

Figure 4.14 Renovation of coastal defense, typical cross section.

Figure 4.15 The pile installation work in progress.

intensity of hurricanes, and the sea level rise. When beach itself has the value in eco-system or for recreation, the construction of huge seawall is not always a good solu-tion. Beach nourishment by detached breakwaters or groins may be useful option if properly undertaken taking into account the use of the coastline and the ecosystem in the surrounding areas. The embedded wall may contribute in such undertakings.

For the emergent protection of residence or facilities behind the beach, embedded wall construction has been carried out. Shown in Figure 4.16 is an example of emer-gency seawall at the town of Lantana, Florida.

Z-shaped steel sheet piles (PZC26), 10.7 m long, were penetrated into very stiff sand layer as shown in the soil profile in Figure 4.17. The project owner specified the press-in piling assisted by simultaneous inner augering in order to minimize the influence on the nearby structures and reduce noise and vibration impacts during construction on restaurant patrons and park visitors.

Figure 4.16 Emergency seawall under construction at the Town of Lantana, Florida.

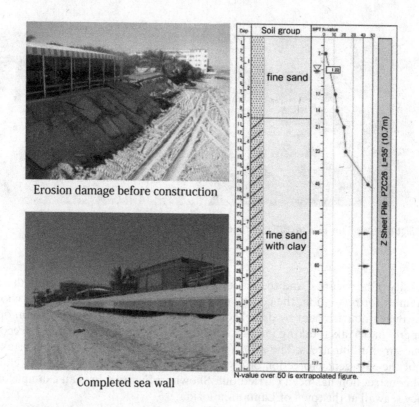

Figure 4.17 Emergency sea wall.

4.5 Seismic reinforcement of various infrastructures by sheet pile wall enclosure

Seismic design standards for various structures have been revised by responsible authorities around 10–20 years interval based on the lessons learned from disasters caused by huge earthquakes. Old structures were constructed to meet the requirements of then-available design standard and are recommended or required to retrofit. Therefore, renovation or upgrading of old structures often accompany the seismic retrofitting as shown earlier in the coastal defense in the previous section and bridges in Bangladesh described in Chapter 3.

The following example is a seismic retrofitting of oil tank in the Japan's coastal industrial areas. By the government ordinance issued in 1994, the improvement of large oil tanks against liquefaction was mandated within a certain transitional deadline. The preferred countermeasures taken to liquefiable foundation soils were ground improvement by grouting, ground water lowering, and steel sheet pile ring method.

Steel sheet pile ring is a continuous embedded wall surrounding the tank foundation as shown in Figure 4.18. The functions of embedded steel sheet pile wall are: restraining the shear deformation of loose soil beneath the tank and reducing the excessive pore pressure generation, preventing lateral flow of foundation soil and local slip failure at the periphery of tank, and shutting off pore pressure propagation to the confined soil even if the soil outside the ring liquefied. Steel sheet pile ring is constructed by installing straight steel sheet piles down to non-liquefiable layer. Especially in densely developed areas, the retrofitting works have to overcome the spatial constraints and the influence of noise and vibration.

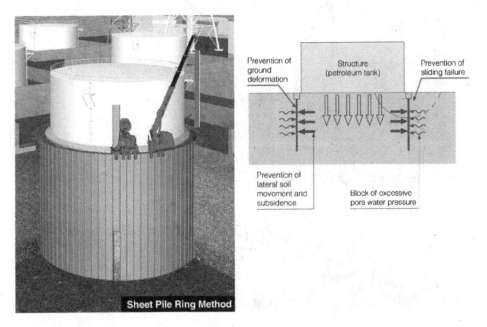

Figure 4.18 Steel sheet pile ring wall for liquefaction countermeasure.

Figure 4.19 is the actual construction of the steel sheet pile ring around a petroleum tank at Shimonoseki, Japan, owned by the Chugoku Electric Power Co., INC. Due to the spatial constraints and the necessity to eliminate the adverse influence of vibration and displacement, Silent Piler was used together with the Non-staging system comprising clamping crane and pile runner.

Sheet pile wall enclosure can be applied to variety of structures above ground or underground as illustrated in Figure 4.20. The floating up of light-weight underground structures such as utility tunnels can be effectively prevented by the enclosure. Road/railway embankments can be seismic retrofitted by means of a pair of sheet pile walls at the toes of embankment.

East Japan Railway Company has been undertaking seismic reinforcement of various facilities against an anticipated earthquake directly beneath the Tokyo

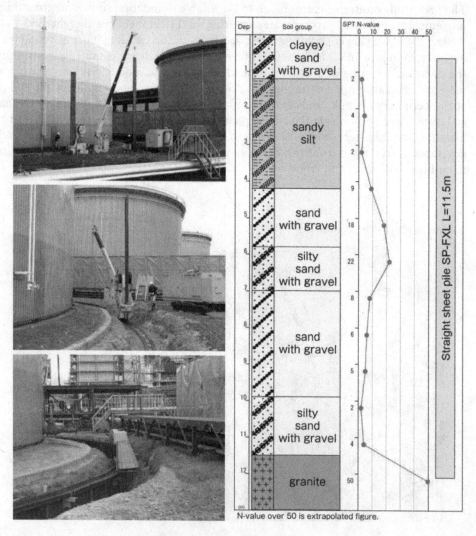

Figure 4.19 Construction of steel sheet pile ring around a petroleum tank.

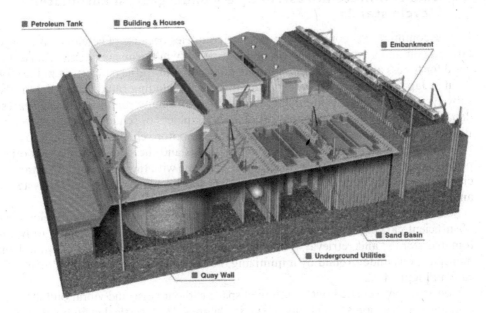

Figure 4.20 Sheet pile wall enclosure for a variety of structures.

metropolitan area which include embankment and cut slopes, bridges, tunnel, and stations. Figure 4.21 is one of such examples; a seismic reinforcement of embankment between Higashi Kanagawa and Ohkuchi of JR Yokohama Line. Double steel sheet pile wall is constructed on toes of embankment and connected each other at upper portion by tie wire to increase its stability and reduce settlement during earthquake.

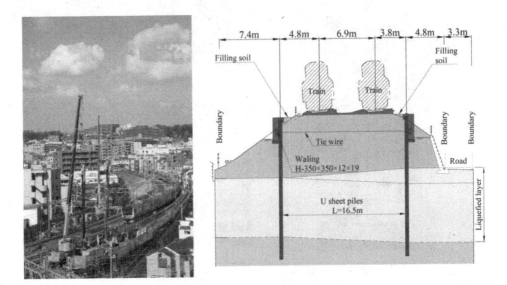

Figure 4.21 Seismic reinforcement of railway embankment.

4.6 Bike commute infrastructure – underground automated bicycle stands

Bicycle commuting is popular in developing countries because of economy. Bike also attracts commuters in developed countries, especially in European cities, because it is good for fitness and also it reduces the air pollution and traffic congestion in the city. Even in Japan, where public transportation system including rapid trains, light rails, subways, and city buses is highly developed, bike commute from home to the nearest departing station and the arriving station to work places or schools is quite common. Bike commute, however, has increased crash with motor vehicles and/or pedestrians. Parked bikes in great number in the public squares and the sidewalks obstruct other people's passages, especially of wheel chairs. To cope with these difficulties, advanced cities are investing in bicycle infrastructures such as bike lanes, bike-sharing systems, and bike parking stands.

Scarcity of available land in densely populated city has prevented the development of sufficient parking space for bicycles. "Eco-Cycle" developed in Japan is a fully automated storage and retrieval system for bicycles, which provides bicycle stands on multiple levels underground by minimizing the use of land space as shown schematically in Figure 4.22.

Construction is carried out in a limited space without noise and vibration following a simple construction sequence as shown in Figure 4.23. A vertical cylindrical shaft is constructed by installing steel sheet piles down to required depth. Basement concrete slab and internal ring beams are installed to complete vertical shaft. Steel sheet piles function to retain surrounding soils and to shut off ground water. A fully automated

Figure 4.22 Eco-Cycle – underground automated bicycle stands.

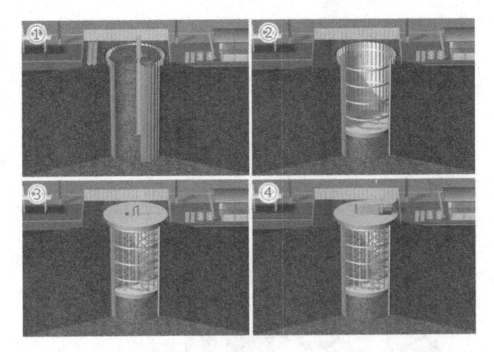

Figure 4.23 Construction sequence of Eco-Cycle.

storage and retrieval system for bicycles is assembled in the shaft. Finally, the ground floor slab and compact bicycle entrance booth is constructed at grade. After completion, valuable space at ground surface can be left open for public use or for other purposes.

Currently available standard model of Eco-Cycle, which accommodates 204 bicycles, requires inside dimensions of the vertical shaft with only 8.1 m diameter and 11.9 m depth. Storage and retrieval of bicycle is computer controlled with the aid of IC (Integrated Circuit) user card and IC tag attached to the bicycle. Average retrieval time is only 13 seconds.

As of August 2018, 52 units of Eco-Cycle have been adopted at 22 locations in Japan, which include the locations near the railway station, city hall, park, shopping district, school, and rent-a-cycle depot. Figure 4.24 shows the construction of Eco-Cycle in Minato-ku, Tokyo.

4.7 Other interesting applications

4.7.1 Preservation of historic structures

Historic structures, especially those made of stone or brick founded on shallow foundations, are vulnerable to vibration and displacement. Adjacent construction such as foundation work and excavation needs to be carefully planned taking the structural characteristics of historic structures into account. Regarding pile installation, the press-in piling with low vibration is superior to the ordinary vibratory or impact hammering methods.

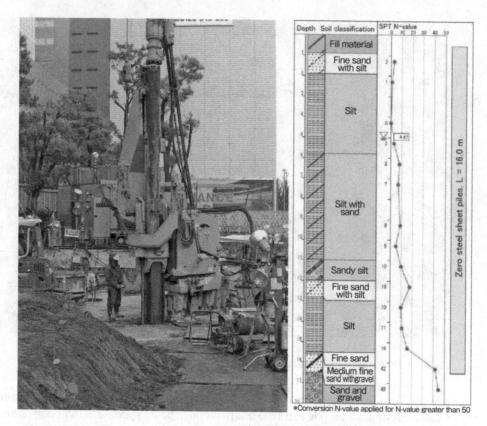

Depth	Soil classification	SPT N-value 0 10 20 30 40 50	
	Fill material		
	Fine sand with silt	3	
	Silt		
		0.6	
		4.97	
	Silt with sand		
	Sandy silt	10	
	Fine sand with silt	10	
		10	
	Silt	11	
		16	
	Fine sand	42	
	Medium fine sand with gravel		
	Sand and gravel	45	

Zero steel sheet piles, L = 16.0 m

*Conversion N-value applied for N-value greater than 50

Figure 4.24 Underground automated bicycle stands in Minato-ku, Tokyo.

San Juan de Ulúa is a fortress overlooking the port of Veracruz, built by the Spanish Empire in the middle of 16th century on a sandy island. The port of Veracruz is one of the busiest ports in Mexico. Due to the expansion of navigation channel and increased number of vessels, the foundation soil of the fortress has experienced erosion. Structures such as ramparts that face the navigation channel were threatened with collapse by settlement and tilting.

To restore the undermined wall foundation and to prevent further erosion, U sheet piles, LX 32 were installed down to 18 m and the foundation soil was reinforced by concrete filling. The protective steel sheet pile wall was constructed on the bay side of the fortress as shown in Figure 4.25.

The requirements for the pile installation work were minimizing vibration to restrain adverse influence on the wall structure and minimizing the space for construction barges and temporary structures to restrain adverse influence on traffic near the fortress. The press-in piling by the Silent Piler was adopted as shown in Figure 4.26.

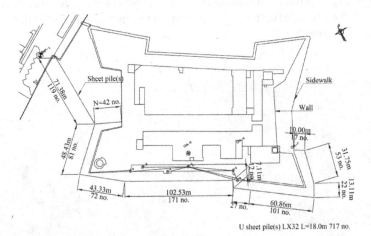

a. Pile layout (Plan)

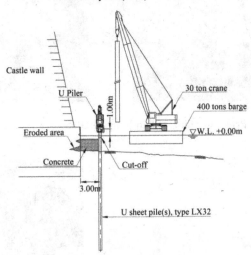

b. Machine layout (Cross-section)

Figure 4.25 Restoration of undermined wall foundation by embedded wall. (a) Pile layout (plan). (b) Machine layout (cross-section).

4.7.2 Protection of residence from secondary disaster near the sink hole

This is an example of emergent stabilization of slope created by sinkhole. The location is Orland, Florida, where bedrock is mostly limestone and susceptible to sinkholes. The photograph on the left-hand side of Figure 4.27 shows the slope close to the apartment covered tentatively by plastic sheet to prevent erosion by rain. To prevent the secondary disaster, permanent retaining wall was constructed by tubular piles with interlocking device, with 900 mm diameter and 15 m length as shown in Figure 4.28. Considering the safety of the building foundation and the unstable ground itself, the press-in piling assisted by simultaneous inner augering was adopted. Nam (2019) and Takuma (2019) provide further information on the mechanisms of sinkhole formation and countermeasures.

Figure 4.26 Press-in piling at San Juan de Ulúa.

Figure 4.27 Protection of Woodhill apartment complex close to the sinkhole.

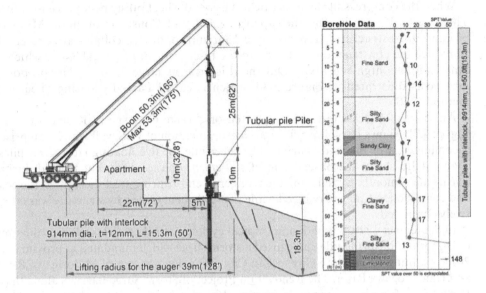

Figure 4.28 The typical cross-section and the bore hole log.

4.7.3 Sheet piling to control settlement due to adjacent construction

Adjacent construction would more or less affect the surrounding environment and nearby structures. Especially the embankment construction on soft clay deposit influences larger extent outside of its right-of-way. Figure 4.29 illustrates such a problem and expected effect of one of the possible countermeasures. The sheet pile wall confines the propagation of additional stresses induced by embankment loading underneath the embankment and protects the nearby structures from the settlement. The initial idea was the construction of continuous sheet pile wall down to the bearing stratum underneath the compressible layer as shown in the figure.

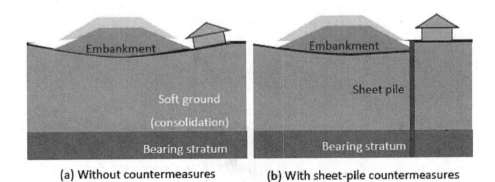

(a) Without countermeasures   **(b) With sheet-pile countermeasures**

Figure 4.29 Problem associated with adjacent construction and countermeasure by sheet piling (Otani, 2017). (a) Without countermeasures and (b) with sheet pile countermeasures.

When the compressible layer is thick, the cost of sheet piling becomes enormous. The Kyushu University and the Japan's Ministry of Construction (now, Ministry of Land, Infrastructure, Transport and Tourism) conducted collaborative research to reduce the construction cost even in the thick deposit of Ariake Clay, which is famous for its high sensitivity and the thickness as large as 40 m. The proposed method is illustrated in Figure 4.30, which is called Partial Floating Sheet-Pile method, PFS.

An in-situ full-scale test was conducted on Ariake Clay site in Kumamoto city. Figure 4.31a shows the soil profile and the specification of PFS wall which comprises a combination of one end-bearing steel sheet pile for five floating steel sheet piles. Figure 4.31b shows the observation of settlement due to embankment load during 4 years and 1 month. By reducing the number of end-bearing steel sheet piles, the cost and construction time are reduced while the settlement of nearby structure was effectively controlled.

In September 1999, a typhoon (a tropical cyclone) attacked Ariake Bay at the time of high water of spring tide and caused a storm surge which inundated approximately 20 ha spanning Kumamoto City to Uto City along downstream of Midorigawa River and Hamadogawa River. The Renovation Project of Storm Surge Barriers along these rivers to reinforce by raising and widening the levees is being undertaken by the Ministry of Land, Infrastructure, Transport and Tourism. Figure 4.32 shows a renovation plan of a levee on the right bank of Hamadogawa River at Hashirigata Machi, Uto City. Shaded area is the additional fill on the existing levee. Prior to the additional filling, two sheet pile walls were installed at both sides of the levee.

Sheet pile wall on the left-hand side is constructed by installing 26.5 m long U sheet pile SP-IVw with 600 mm effective width which is expected to function as a cantilevered wall to maintain the stability of levee during additional filling. The sheet pile wall on the right-hand side is the PFS wall to control settlement outside the levee's foot print. The PFS wall comprises a combination of an end-bearing steel sheet pile about 40 m long for nine floating steel sheet piles about 27 m long. By using Hat-shaped steel sheet pile SP-25H with 900 mm effective width, number of piles were reduced in comparison with ordinary U sheet pile and resulted in reduced construction cost and time.

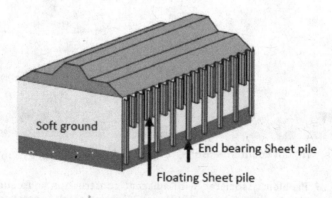

Figure 4.30 Partial floating sheet pile method. (Otani, 2017).

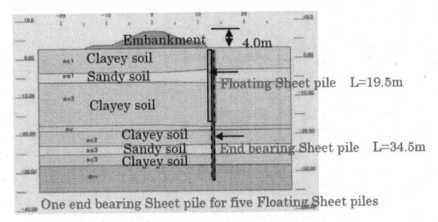

a. Soil profile and PFS wall specifications

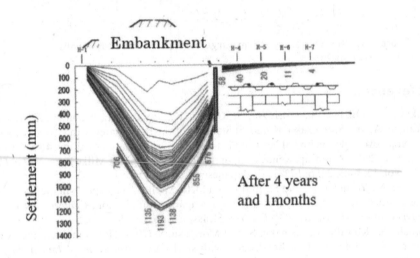

b. Settlement observation

Figure 4.31 An in-situ full-scale embankment to investigate PFS wall (Otani, 2017). (a) Soil profile and PFS wall specifications. (b) Settlement observation.

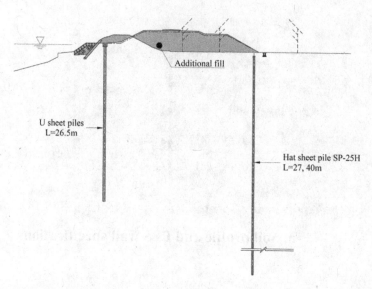

U sheet piles
L=26.5m

Additional fill

Hat sheet pile SP-25H
L=27, 40m

Figure 4.32 Renovation of storm surge barrier by embedded walls.

References

Arikawa, T., Oikawa, S., Moriyasu, S., Okada, K., Tanaka, R., Mizutani, T., Kikuchi, Y., Yahiro, A. and Shimosako, K. (2015) Stability of the breakwater with steel pipe piles under tsunami overflow, Technical Note of Port and Airport Research Institute, 46 p. (in Japanese)

Hiiragi, S. (2006) Zoom-up: Renovation of navigation channel at Miike Port, *Nikkei Construction* Vol. 12, Issue 22, pp. 72–75 (in Japanese).

Ishihara, Y., Yasuoka, H. and Shintaku, S. (2020) Application of press-in method to Coastal Levees in Kochi Coast as countermeasures against liquefaction, *Geotechnical Engineering Journal of the SEAGS & AGSSEA*, Vol. 51, Issue 1, pp. 79–88.

Kikuchi, Y., Kawabe, S., Taenaka, S. and Moriyasu, S. (2015) Horizontal loading experiment on reinforced gravity type breakwater with steel walls, *Proceeding of 15th Asian Regional Conference on Soil Mechanics and Geotechnical Engineering*, pp. 1268–1271.

Kimura, Y. (2012) Renovation of sea wall, *Construction Machinery and Equipment*, Vol. 2, pp. 7–12 (in Japanese).

Moriyasu, S., Tanaka, R., Oikawa, S., Tsuji, M., Taenaka, S., Kubota, K. and Harata, N. (2016) Development of new type of breakwater reinforced with steel piles against a huge tsunami, Nippon Steel & Sumitomo Metal Technical Report No. 113, December 2016, pp. 64–70

Nam, B.H. (2019) Detection and geotechnical characterization of sinkhole: Central Florida case study, *IPA Newsletter*, Vol. 4, Issue 3, pp. 9–14.

Otani, J. (2017) A new steel sheet pile method for countermeasures against the settlement of embankment on soft ground –development of PFS method-, *IPA Newsletter*, Vol. 2, Issue 3, pp. 8–10.

Takuma, T. (2019) Mitigation of sinkhole with press-in Piles, *IPA Newsletter*, Vol. 4, Issue 3, pp. 15–18.

Chapter 5

Responses of piles installed by the press-in piling

5.1 Introduction

Practical applications of press-in piling have been widely accumulated since the time of invention of Silent Piler. As summarized in the Appendix (List of related publications), a number of research papers have been published over these 20 years or so. This chapter describes geotechnical aspects of the press-in technology, introducing responses of piles installed by press-in piling, mainly based on the research outcomes obtained from the research collaborations between the University of Cambridge and GIKEN LTD. for more than 20 years. Issues to be presented in this chapter include "Ground vibration and noise during pile installation", "Ground responses and pile resistance during pile installation", and "Performance of a single pile installed by press-in piling".

5.2 Ground vibration and noise during pile installation

5.2.1 Introduction

As was described in Chapter 2, the Silent Piler was originally developed for minimum environmental impact during piling operations with respect to ground vibration and noise. One of the key features of the press-in piling is environmental friendliness. The ground vibration and noise caused during the installation of pre-fabricated piles by dynamic methods can lead to human disturbance and structural damage. Stringent regulations now virtually preclude the installation of piles by dynamic methods in urban environments in Japan, Europe, and other countries. The performances of the Silent Piler with respect to noise and vibration have been studied by field monitoring.

5.2.2 Piling-induced noise pollution

Noise levels are expressed in decibels and are derived from the fluctuating air pressure.

$$\text{Air pressure, p } (\mu\text{Pa}) = 20 \times 10^{(\text{dB}/20)} \tag{5.1}$$

The perceived noise level of a source, L_{source}, reduces with distance from the source and the attenuated noise level which neighbors will experience ($L_{equivalent}$) can be expressed as in Equation 5.2.

$$L_{equivalent} = L_{source} - 20\log{(r)} - 8 \tag{5.2}$$

where r is the horizontal distance from the piling operation in meters.

White et al. (2002) presented the variations of noise level with distance from various types of piling operation. Accumulated field data are plotted in Figure 5.1, which proves that the noise level caused by press-in piling is well below acceptable levels specified by the regulations (e.g., BS 5228).

5.2.3 Piling-induced ground vibration

Ground vibrations are usually quantified by the peak velocity of particles in the ground as they are disturbed by the passing wave (peak particle velocity – PPV). PPV is the preferred measure of the damage potential of a vibration. Instantaneous particle velocity consists of three orthogonal components which are usually measured independently using a triaxial geophone. The most commonly used definition of peak particle velocity is the simulated resultant PPV.

Figure 5.2 presents measured ground vibrations during dynamic piling (Head & Jardine, 1992). Significant scatter is observed in the data, in particular, the zone over about 10 m away from the vibration source for the case of diesel hammer operation.

White et al. (2002) reported interesting field monitoring data of ground vibration during press-in piling at two sites: Site 1: New Orleans, USA and Site 2: Utrecht, the Netherlands. The data at Site 2 provides a direct comparison of the ground vibrations

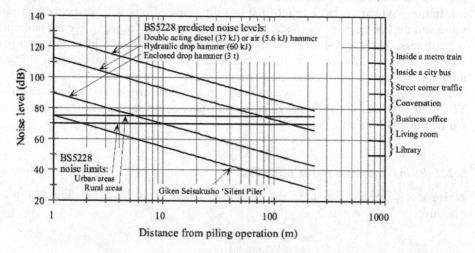

Figure 5.1 Variation of noise level with distance from piling operation (White et al., 2002).

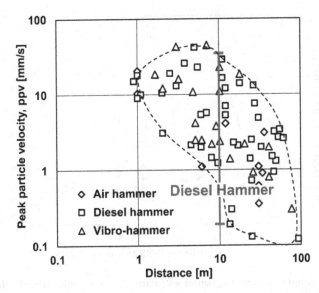

Figure 5.2 Data of measured ground vibration: modified from Head & Jardine (1992).

created by three different methods of piling. They described the construction sequence as follows.

A wall of 500 mm U-shaped steel sheet piles was installed beside Meester Tripkade, Utrecht, Holland as temporary works during the widening of the Utrecht to Blaauwkapel railway line during July 1992. Vibration monitoring was carried out by Dutch Railways to assess any disturbance to the foundations of nearby properties. Triaxial geophones were attached to the foundations of three houses 150 mm above ground level, located 7.15 m from the piling line. Local regulation recommended that ground vibrations be limited to 3 mm/s around residential buildings.

Construction of the wall commenced using a diesel hammer. However, the measured resultant PPV was 15.2 mm/s, significantly exceeding the limit. A subsequent pile was installed using a vibratory method, with a resultant PPV of 8.3 mm/s being recorded. A Silent Piler was then brought onto site and the remaining piles were installed by the press-in method. Peak particle velocities in the range of 0.3–0.7 mm/s were recorded during this stage of the construction. The influence of soil conditions and pile type can be ignored since identical piles were being installed in identical soil.

Figure 5.3 shows that replacing the dynamic piling equipment with a press-in pile driver leads to a 10–50 times reduction in ground vibrations. From this limited data base, they proposed the best-fit line as

$$V \text{ press} - \text{in (mm/s)} = 7 / r \text{ (m)} \tag{5.3}$$

Rockhill et al. (2003) gathered ten sets of data of ground vibration during press-in piling as are listed in Table 5.1.

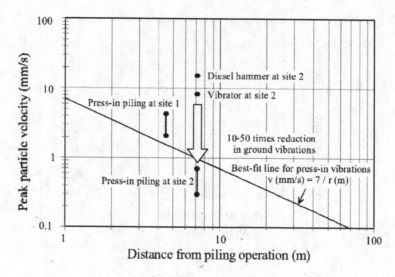

Figure 5.3 Improvement of ground vibration by Introducing Silent Piler (White et al., 2002).

Table 5.1 Descriptions of test site (Rockhill et al., 2003)

Test	Location	Date	Soil properties	Piler type	Pile type
1	Takasu test site, Kochi, Japan	July 2002	Made ground overlying silty sand	Giken Super Auto 75	0.4 m × 6.5 m sheet piles
2	Takasu test site, Kochi, Japan	July 2002	Made ground overlying silty sand	Giken NT 150	0.1 m diameter, 8 m tubular piles
3	Takasu test site, Kochi, Japan	July 2002	Made ground overlying silty sand	Diesel generator (1,800 rpm)	N/A
4	Othu Funaire, Kochi, Japan	July 2002	Loose, stony fill overlying	Giken Super Auto 150	0.4 m × 10 m sheet piles
5	Othu Funaire, Kochi, Japan	July 2002	Loose, stony fill overlying	Type SS-40L low-amp high-freq vibro-ha miner	0.4 m × 12 m sheet piles
6	Tosashi, Kochi, Japan	July 2002	Loose, stony fill	Giken Super Auto 100 (water jetting @ 7 MPa)	0.4 m × 14.5 m sheet piles
7	Atago, Kochi, Japan	July 2002	Made ground overlying silty clay	Giken Super Crush 100M (auger)	0.4 m × 8 m sheet piles

(Continued)

Table 5.1 (Continued) Descriptions of test site (Rockhill et al., 2003)

Test	Location	Date	Soil properties	Piler type	Pile type
8	Iriake, Kochi, Japan	July 2002	Rocky made ground	Giken Super Auto 75	0.4 m × 6 m sheet piles
9	Westbourne Grove, London	January 2003	Rubble fill over soft clay and London Clay	Giken Super Auto UPI50 (water jetting for lubrication)	0.6 m × 12 m sheet piles
10	Norway	Autumn 1998	Silt, sand, and clay	Giken ZPI50	0.6 m × 15 m sheet piles

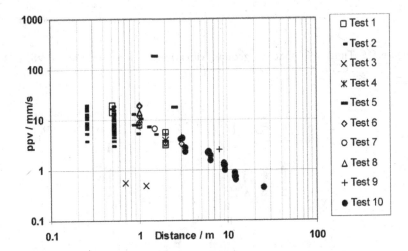

Figure 5.4 Data of ground vibrations on ten test sites (Rockhill et al., 2003).

The results are presented in Figure 5.4. Although there are notable scatters near the vicinity of the pile operation up to 1 m away from the piling operation, it seems that all the data linearly decay with the logarithm of the distance within a narrow range of variation.

As was described in Chapter 2, the press-in piling operation involves a cycle of operation: installation and walking. The full-depth installation is made by a number of successive jacking strokes. It is interesting to identify which stage of pile installation causes a major source of ground vibration, which could lead to further improvements of operation on site as well as mechanical design.

Figure 5.5 shows the time record of ground vibrations during installation of a sheet pile using a Silent Piler (White & Deeks, 2007). Negligible vibrations are recorded while the pile is moving, but transient pulses are evident. Looking at a single jacking stroke, the largest of these events corresponds to the opening of the chuck that grips the piles at A and from the impulse of hydraulic force at the start of the next jacking stroke at D.

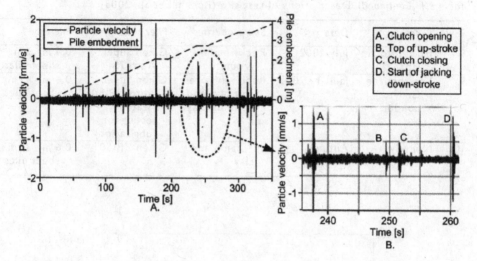

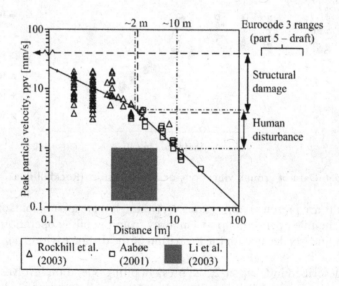

Figure 5.5 Time sequence of ground vibration by a Silent Piler (White & Deeks, 2007).

Figure 5.6 Collected data sets of ground vibration (White & Deeks, 2007).

Ground vibration measurements near to pile jacking operations have also been reported by Aabøe (2011), Li et al. (2003), and Rockhill et al. (2003). White & Deeks (2007) gathered these data and presented them as is seen in Figure 5.6.

There are a number of different empirical predictions and all take the form of a power law as is given in Equation 5.4.

$$PPV = C\{ w^{0.5}/r\}^n \tag{5.4}$$

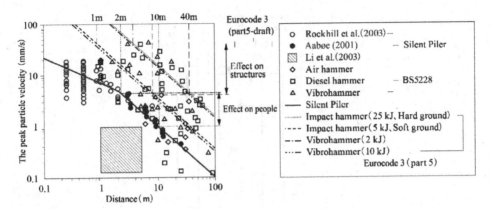

Figure 5.7 Collected data of ground vibration for various machines (IPA, 2016).

where w: hammer energy per blow or cycle (J/blow or J/cycle)
   r: horizontal distance from the piling operation in meters
   PPV: the predicted peak particle velocity in mm/s
   C: empirical parameter related to soil and hammer type.
The value of C typically varies from 0.5 to 1 in the various standards and studies.
   Rockhill et al. (2003) proposed a bilinear fit as a simple predictor for ground vibrations induced by press-in piling as below.

$$PPV \text{ (mm/s)} = \min\left\{ 7.7/r^{0.5}, 10.4/r \right\}(r \text{ in meters}) \tag{5.5}$$

Figure 5.7 is a summary chart of the ground vibration, combining Figures 5.2 and 5.6. It is confirmed from the figure that the Silent Piler causes much smaller ground vibration at least within about 10 m away from the piling operation.

5.3 Ground responses and pile resistance during pile installation

5.3.1 Introduction

The installation method of a pile determines the stress state and level of disturbance of the surrounding soil, which in turn has a significant influence on the subsequent behavior of the resulting foundation pile. As was described in Chapter 2, the pile installation process by the Silent Piler typically includes the three modes of penetration: (i) monotonic penetration, (ii) repeated penetration and extraction, and (iii) rotary penetration.
   The walk-on-pile-type press-in piling involves the loading and unloading process during the construction sequence of pile installation as the construction proceeds as is illustrated in Figure 5.8, since utilizing the previously installed piles as a source of reaction force is the basic principle of the walk-on-pile-type press-in piling. More precisely, a pile experiences (i) downward load at the termination of its own press-in

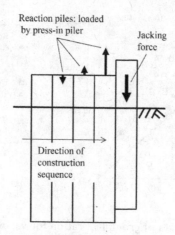

Figure 5.8 Construction sequence of pile installation by Silent Piler.

process, (ii) upward load when it is used as the first or second reaction pile, and (iii) downward load when it is used as the third or fourth reaction pile.

The press-in piles are often subjected to the repeated penetration and extraction process on site for various reasons, earlier touched upon as the up and down motion in Chapter 2. Machine operators often aim to reduce the required press-in force in order to counter the tendency for sheet piles to deviate from vertical during hard driving. The option of the repeated penetration and extraction process is also adopted when adjusting the pile position precisely to the design alignment and also when the machine capacity is insufficient to push the pile in the ground, for example, when the pile encounters an expected hard layer of soil. If machine operators encounter these situations, it is common practice for them to reverse the displacement of the pile by extracting the pile for a given stroke following which the pile is reinserted. The process is sometimes termed as surging.

Rotary penetration is another installation method within the family of the Silent Piler, in which torque and axial force are applied simultaneously to install steel tubular piles (hereinafter called tubular piles) by rotating them into the ground. The machine called Gyro Piler adopts a rotary penetration mechanism for installing a tubular pile with cutting bits into the ground, even in rock or in locations having obstacles such as existing reinforced concrete structures. The Gyro Piler is equally based on the basic principle of the walk-on-pile-type press-in piling, utilizing the previously installed piles as a source of reaction force, so that a cyclic loading and unloading process is also involved. In general, by selecting an appropriate number of cutting bits the Gyro Piler has a sufficient penetration capacity for a given site condition and also the circular geometry of the tubular pile ensures the verticality and precise positioning of piles.

5.3.2 Ground response mechanism under monotonic penetration

5.3.2.1 Sheet pile

Sheet pile installation into the ground may be a complicated process due to a possible multi-layered system of the ground, due to various cross-sectional geometries of the sheet piles, and due to possible plugging in some stages of penetration. To gain an insight of the ground responses during sheet pile installation, detailed observation of ground response is of use in a well-controlled laboratory test with a simple geometry of the sheet pile. An idealized model of sheet pile installation may be a plate penetration in the plane strain condition.

White (2002) conducted a series of model penetration test of a metal plate (model sheet pile) in two different dry sands: carbonate sand (DBS, Dog's Bay Sand) and silica sand (LBS, Leighton Buzzard Sand) in the plane strain condition. The flat-ended pile of breadth 32.2 mm was monotonically penetrated with a penetration rate of 1 mm/min by a machine screw-driven actuator. Detailed measurements of soil movement around the plate were made through observation windows by the combined techniques of digital photography, particle image velocimetry (PIV), and close-range photogrammetry with high precision.

The classical bearing capacity theory of deep foundation by Terzaghi for the base resistance in the two-dimensional situation assumed the Prandtle-type failure mechanism with a surcharge pressure due to overburden pressure and is expressed as

$$q_b = \gamma BN_\gamma/2 + \gamma zN_q \tag{5.6}$$

where B: breadth of the footing

z: embedded depth (penetration depth)

N_γ, N_q: bearing capacity factors, exponential functions of friction angle.

Equation 5.6 implies that the relationship between base resistance and embedment depth exhibits a bilinear relationship. After reaching the full mobilization of the first term in Equation 5.6, the base resistance linearly increases with depth due to the second term.

Figure 5.9 presents the curves of pile base resistance q_b with depth during monotonic installation for the carbonate sand (DBS) and the silica sand (LBS), each carried out under vertical surcharge on the top surface of 50 kPa equivalent roughly to an additional depth $z_0 \approx 3$ m (White & Bolton, 2004). Because of the surcharge, it would have been expected that the second term in Equation 5.6 would have risen only about 10% at a depth of 0.3 m, whereas the curves for the silica LBS show a lager rate of increase, attributed to boundary effects. The very much smaller penetration resistance of the carbonate DBS, notwithstanding its higher angle of friction in conventional triaxial tests at low confining pressures, is notable. This is attributed to the effects of greater soil crushability in the case of the DBS which eliminates the effects of dilatancy even at moderate stress levels, reducing the angle of shearing accordingly and therefore strongly reducing the bearing capacity factors. However, the applicability of Equation 5.6 was further challenged by observing the mechanism of soil–pile interaction.

The most straightforward illustration of the penetration mechanism is the displacement field around the pile tip. Figure 5.10 shows the displacement field captured by PIV, indicating that there is no evidence of formation of a bearing capacity-type mechanism (shown in dotted lines) in which the soil flows along streamlines curving from below the pile tip to the upward direction on either side of the shaft.

Figure 5.11 shows the full displacement trajectories of two soil elements tracked through the tests of the carbonate sand and the silica sand. The coordinate origin is located on the centerline of the pile and level with the pile tip at the end of installation.

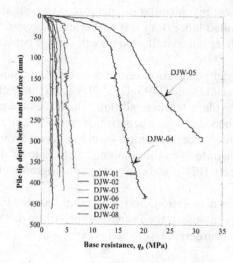

Calibration chamber test series

Test identity	Sand type	Initial voids ratio, e_0	Relative density (%)	Pile breadth (mm)	Pile tip configuration
DJW-01	DBS	1.30	64	32.2	Flat-ended
DJW-02	DBS	1.48	44	32.2	Flat-ended
DJW-03	DBS	1.24	71	32.2	Flat-ended
DJW-04	LBS	0.70	34	32.2	Flat-ended
DJW-05	LBS	0.64	55	32.2	Flat-ended
DJW-06	DBS	1.12	84	32.2	Flat-ended
DJW-07	DBS	1.47	45	32.2	Driving shoe
DJW-08	DBS	1.46	46	16.1	Flat-ended

Index properties of test sands

Sand	Mineralogy	G_s	D_{50} (mm)	e_{max}	e_{min}
DBS	Calcium carbonate	2.75	0.44	1.87	0.98
LBS	Silica	2.65	0.84	0.80	0.51

Figure 5.9 Load intensity versus settlement curves (White, 2002).

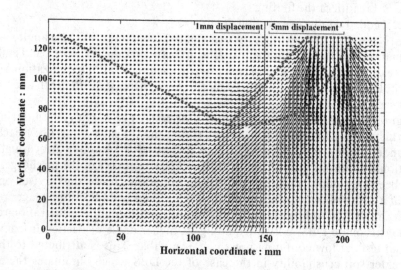

Figure 5.10 Displacement field (White, 2002).

As the pile approaches, the movement is generally downward, with the soil element trajectory curving toward the horizontal as the pile passes. A notable feature of Figure 5.11 is the tail at the end of each trajectory. After the pile tip has passed the soil element (i.e. $h > 0$, where h is the vertical distance of a soil element above the pile tip), the soil relaxes back toward the pile shaft.

Figure 5.12 presents displacement trajectories of a single column near the pile. It is notable that the trajectories show no systematic variation with depth.

Based on the detailed measurements of strain and rotation paths, White & Bolton (2004) schematically presented the mechanism of the penetration by showing two

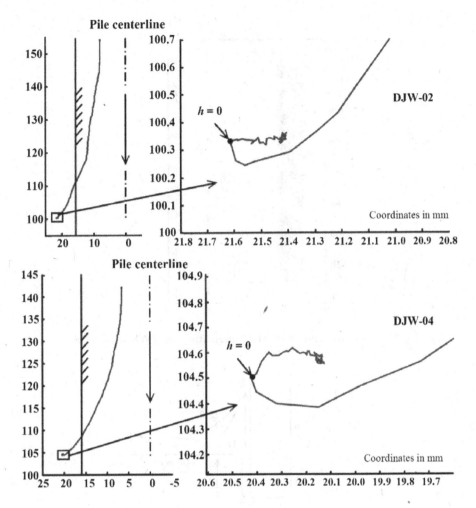

Figure 5.11 Soil movement associated with pile penetration (White, 2002).

streamlines of soil flow as is presented in Figure 5.13. Soil that starts beneath the path of pile flows close to the nose cone formed immediately beneath the pile tip and forms the interface layer adjacent to the pile shaft, following streamline ABC. Soil that was initially just outside the path of the pile follows streamline DEF. The ultimate volumetric strain of sand elements following streamlines ABC and DEF reveals the variation of final density with offset from the pile shaft. This variation is shown schematically at the top of Figure 5.13. Adjacent to the pile shaft, the soil has become denser, following irrecoverable volume change in the zone of high stress and shear strain around the nose cone.

Figure 5.14 summarizes the observed deformation mechanism adjacent to the model pile. This mechanism links the kinematic observation of a contractile interface zone to the degradation of shaft friction close to the pile tip. High horizontal stress is created as soil is compressed laterally along streamline XY. As the soil continues along

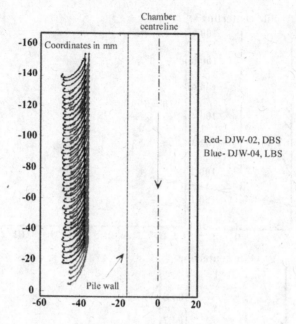

Figure 5.12 Soil movements near the penetrating pile (White, 2002).

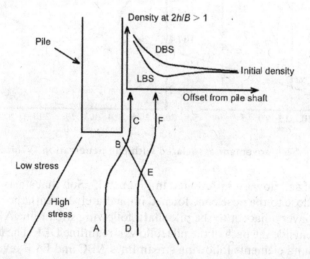

Figure 5.13 Soil movement and volumetric changes near pile tip (White & Bolton, 2004).

streamline YZ, the crushed and disturbed interface zone immediately adjacent to the pile (zone B in Figure 5.14) contracts with continued shearing at the pile–soil interface. The stiff unloading response of the over consolidated soil in the far field is represented by a stiff spring (zone A). This spring is fixed in the far field and exerts horizontal stress on the pile shaft. As separation h increases, zone B contracts and this spring unloads,

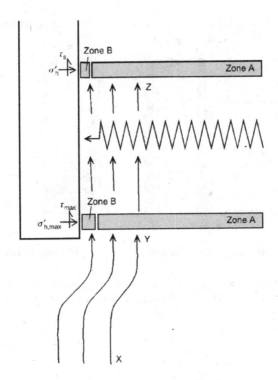

Figure 5.14 Deformation mechanism of plate penetration (White & Bolton, 2004).

reducing the shaft friction on the pile. This phenomenon is well known, as represented by the very low ratio of perhaps 1% between the sleeve friction and cone penetration resistance of cone penetration tests (CPTs) in sand.

5.3.2.2 Tubular pile

Plugging may occur in any concave part of a pile cross section such as the interior of tubular piles, H steel piles, and some types of sheet piles. When installing an open-ended tubular pile, the plugging phenomenon is often observed, especially in sands. Pile base condition whether the base is plugged or unplugged governs the responses of the pile during penetration. Figure 5.15 illustrates the plugged and unplugged modes of penetration.

During unplugged penetration, the pile moves downward relative to the internal soil column. Penetration is resisted by shaft friction on the inside (Q_{si}) and outside of the pile (Q_{so}) and the base resistance on the annulus of pile wall (Q_w). During plugged penetration, the internal soil column is dragged downward, and the pile assumes the characteristic of a closed-ended pile. Penetration is resisted by shaft friction on the outside of the shaft (Q_{so}) and by base resistance both on the pile wall (Q_w) and the soil plug (Q_p).

By considering the static equilibrium of plugged and unplugged penetration, the following equations are derived.

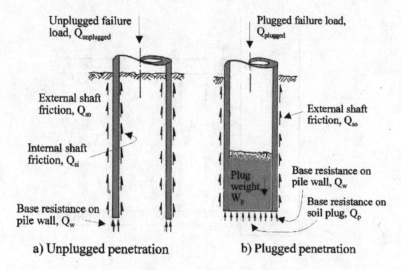

a) Unplugged penetration   b) Plugged penetration

Figure 5.15 Plugged and unplugged mode of penetration (White et al., 2000). (a) Unplugged and (b) plugged modes.

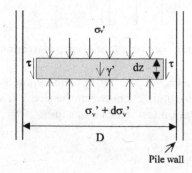

Figure 5.16 Force equilibrium inside the pipe (White et al., 2000).

$$Q_{unplugged} = Q_{so} + Q_{si} + Q_w \tag{5.7}$$

$$Q_{plugged} = Q_{so} + Q_w + Q_p - W_p \tag{5.8}$$

where W_p: the plug weight.

The lower bound of these two mechanisms will govern the press-in force and hence the mode of penetration. Plugging will occur if $Q_{si} + W_p > Q_p$.

Local shaft resistance is generally predicted using the Coulomb equation. To find the total external shaft friction (Q_{so}), Equation 5.9 is integrated over the pile surface area. It is often assumed that the earth pressure coefficient, K, and pile–soil friction, δ, remain constant along the length of the pile with limiting values of the local shaft friction applying.

$$\tau_s = \sigma'_h \tan\delta = K\sigma'_v \tan\delta \tag{5.9}$$

$$Q_{so} = \int \tau_s \pi D \, dz \tag{5.10}$$

Referring to Figure 5.16, equilibrium analysis of a horizontal slice of soil shows that vertical stress inside the tube increases exponentially with depth.

$$d\sigma'_v /dz = \gamma' + 4K\sigma'_v \tan\delta/D \tag{5.11}$$

$$\sigma'_v = \gamma' D\left(e^{4Kh\tan\delta/D} - 1\right)/4K\tan\delta \tag{5.12}$$

The total internal shaft friction can be found by integrating Equation 5.11 over the height of the soil column, h, and considering the overall equilibrium of the column.

$$Q_{si} = \sigma'_v \pi D^2/4 - W_p = \left(\gamma' D^3 \pi/16K\tan\delta\right)\left(e^{4Kh\tan\delta/D} - 1\right) - \gamma' h\pi D^2/4 \tag{5.13}$$

Equation 5.13 can adopt the following recommended design variables for medium dense sand in a steel tube, $K = 0.8$ and $\delta = 25°$.

The internal friction drags down on the column of soil inside the pile, increasing the vertical stress within it. This in turn increases the lateral stress inside the pile, which therefore increases the side friction. The positive feedback process causes stresses inside the soil column to increase exponentially. The nature of the inevitable exponential build-up of internal friction inside a hollow pile should be understood by all engineers concerned with press-in operations. The driving force rises increasingly sharply due to this internal friction, even while the central soil body is being cored while remaining in place. The onset of plugging is not the onset of hard driving. Rather it is the relief of even harder driving by the agency of the newly preferred mechanism of plugged penetration.

Figure 5.17 shows predicted driving load versus embedded depth at a typical site.

White et al. (2000) reported field observations of plugging phenomena, where two test piles of outer diameter of 318.5 and 162.5 mm were installed in the ground profile shown in Figure 5.18. The inside soil column length was measured. Figure 5.19 shows the measured profiles of driving load and soil column length with embedded depth observed in the field test, indicating the multiple transitions from unplugged mode to plugged mode can occur in layered soil.

5.3.3 Pile resistance during repeated penetration and extraction

Operators of press-in machines often make use of large cycles of vertical displacement to reduce the required press-in force. Cyclic loading of each soil element adjacent to the pile shaft is analogous to the response observed in a cyclic interface shear box. Cyclic interface shear of sand leads to gradual densification of the shear zone, even in a sample originally in a dense state.

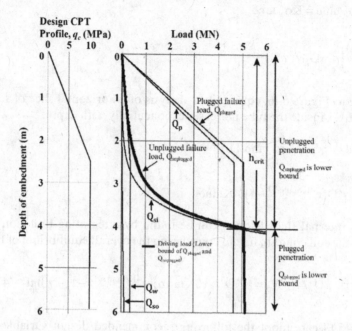

Figure 5.17 Predicted driving load versus embedded depth for plugged and unplugged pile (White et al., 2000).

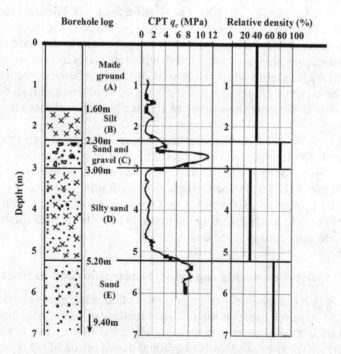

Figure 5.18 Soil profile of field test site (White et al., 2000).

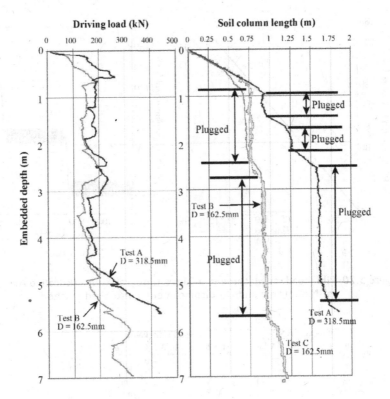

Figure 5.19 Multiple transition from unplugged form to plugged form observed in the field test (White et al., 2000).

When the pile is axially cycled, the shaft friction decreases. The decrease is linked to soil densification at the soil–pile interface which leads soil grains to rearrange, reducing the horizontal effective stress.

Burali d'Arezzo (2015) conducted a series of centrifuge tests, looking into effect of installation method on pile installation load, including displacement-controlled installation and surging installation. Figure 5.20 gives the installation procedure for monotonic (a) and displacement-controlled method (b), plotting pile head displacement against time.

Figure 5.21 presents the centrifuge test data of a closed-ended cylindrical model rough pile installed in a dense dry sand, comparing the monotonic loading with the repeated penetration and extraction in terms of base and shaft resistances with penetration depth. The cycling at constant depth results in a stiffer response and larger base resistance. Further penetration initially causes a slight reduction in base resistance, which then gradually approaches the monotonic curve of base load increasing with depth. The results also show that the repeated penetration and extraction at a constant depth causes a reduction of the shaft resistance compared to a monotonic installation, probably due to friction fatigue mechanism.

The reason for the initially stiffer response and larger base resistance in dense sand may be explained by the formation of a cavity at the base when the pile is extracted,

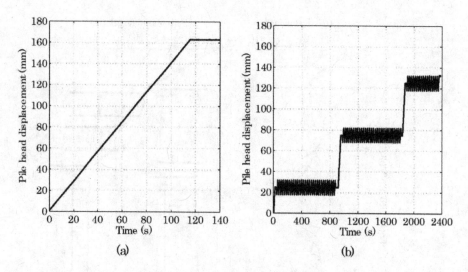

Figure 5.20 Pile head displacement against time relations (Burali d'Arezzo, 2015). (a) Monotonic and (b) displacement-controlled method.

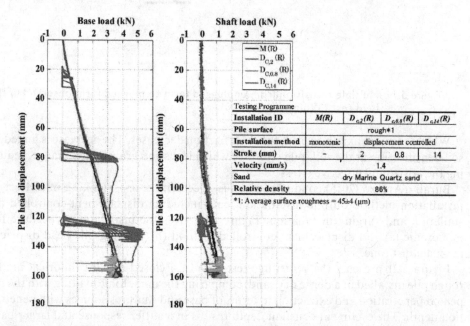

Figure 5.21 Effect of surging observed in centrifuge tests (Burali d'Arezzo, 2015).

with sand that enters the cavity being compacted when the pile is re-penetrated. Post-test observations have shown the presence of crushed grains at the pile tip at the location where the pile was installed. Following grain crushing, soil flows around the pile shoulder and stresses reduce significantly. When the pile is uplifted, the crushed grains will remain at the depth of first uplift while a cavity forms under the pile base.

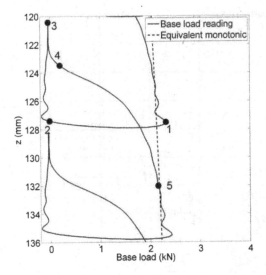

Figure 5.22 Load cycles of surging (Burali d'Arezzo, 2015).

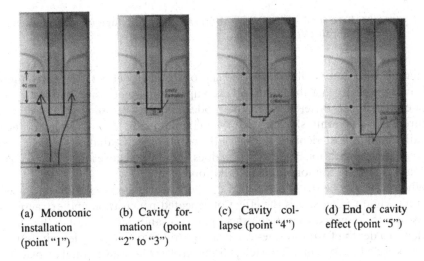

| (a) Monotonic installation (point "1") | (b) Cavity formation (point "2" to "3") | (c) Cavity collapse (point "4") | (d) End of cavity effect (point "5") |

Figure 5.23 Formation of cavity during surging identified by X-ray radiography (Burali d'Arezzo, 2015). (a) Monotonic installation (point "1"), (b) cavity formation (points "2" to "3"), (c) cavity collapse (point "4"), and (d) end of cavity effect (point "5").

Burali d'Arezzo (2015) observed the cavity formation process relative to the penetration depth relationship during the process of penetration and extraction in plane strain chamber tests at 1 g, where a rectangular bar of 12 mm width was penetrated. Figure 5.22 shows the base load changing with the installation depth during two particular cycles of penetration and extraction, where X-ray radiographs were taken at the points from 1 to 5 as shown in Figure 5.23. Two figures identify the cavity

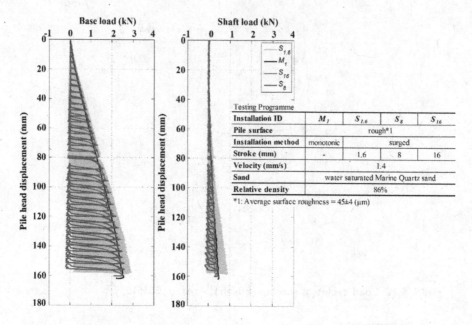

Figure 5.24 Base load —settlement observed in a centrifuge test on a saturated sand (Burali d'Arezzo, 2015).

formation and cavity collapse and the end of the cavity effect. The surrounding soil falls into the cavity and the fallen soil is compacted by successive loading.

The phenomena differed when the model pile was installed in a saturated soil. Figure 5.24 is the result, showing that the maximum base load at the end of the surged installation is almost identical to that of the monotonic installation. This suggests that the repeated penetration and extraction does not always cause an additional base load increase compared to monotonic installation.

White & Deeks (2007) showed field data of sheet pile installation on the effect of surging. Figure 5.25a shows the effect of repeated penetration and extraction on the press-in force required to install a sheet pile at a sandy site. In test 4, the pile was installed to a depth of 10 m by one-way jacking strokes of ~700 mm length. In test 5, the pile was cycled by 350 mm after each jacking stroke. These additional displacement cycles reduced the jacking force by 50% at the final embedded depth. At a different site, with clay layers, the same repeated penetration and extraction surging procedure led to a smaller reduction in jacking force of only 10%–20% as is shown in Figure 5.25b.

This comparison illustrates that surging is less effective in clay since friction fatigue is reduced. In clay, the pile–soil interface shear zone is thinner than for sand, so can undergo less contraction. This strong influence of loading cycles on friction fatigue has implications for the design of jacked piles. Compared to driven piles which are installed by many hundreds of two-way cycles, jacked pile installation involves 10–100 cycles of loading.

Ishihara et al. (2011) provided field data on the effect of surging of a closed-ended steel tubular pile, of which soil profile obtained from CPT is given in Figure 5.26.

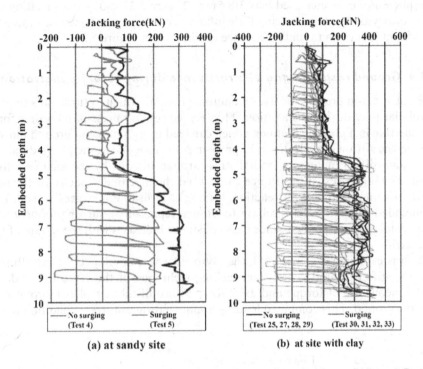

Figure 5.25 Effect of surging on jacking force observed in a field test (White & Deeks, 2007). (a) At sandy site and (b) at site with clay.

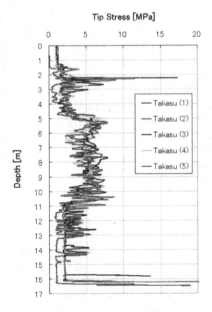

Figure 5.26 Soil profile of the test site obtained from CPT results (Ishihara et al., 2011).

The pile outer diameter used was 318.5 mm. Figure 5.27 shows the variation of base resistance with depth, indicating little influence of surging on base resistance, while a significant reduction in shaft resistance is observed in Figure 5.28.

5.3.4 Ground response and pile resistance during rotary penetration

The pile press-in process involves monotonic loading which in turn leads to plugging of tubular piles during installation. Thin-walled tubular press-in piles therefore tend to penetrate in a plugged manner, which can lead to (i) significant ground movements around the advancing pile, (ii) high press-in resistance due to the plugged base resistance, and (iii) an increase in radial stress around the pile by the high soil displacement. The high press-in resistance can exceed the machine capacity or the negative shaft resistance available for reaction, leading to refusal. For larger diameter-plugged open-ended tubular piles, the base resistance represents a larger proportion of the total installation resistance, since base resistance increases with the square of the pile diameter.

As was explained in Chapter 2, the Gyro Piler was developed for installing a pile into the stiffer ground. The rotary-jacking technique of the Gyro Piler adopts the simultaneous use of torque and axial force to install tubular piles by rotating them into the ground. This combined loading applies a different stress condition to the soil

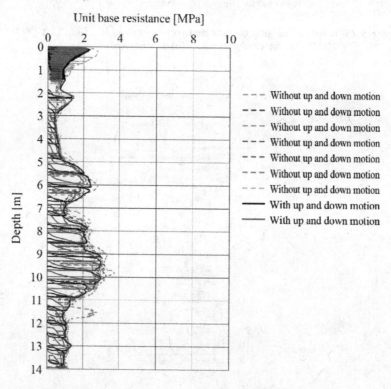

Figure 5.27 Effect of surging on base resistance (Ishihara et al., 2011).

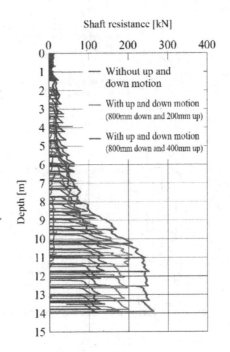

Shaft resistance [kN]

Figure 5.28 Effect of surging on shaft resistance (Ishihara et al., 2011).

around and within the pile, changing the base and shaft resistance compared to purely vertical loading.

Deeks (2008) extensively studied rotary-jacking piling, both from an experimental and theoretical point of view. Figure 5.29 illustrates the soil resistance around the pile, where the axial load at the pile head, on the pile shaft, and at the pile base are denoted by V, Q_s, and Q_b, torsional loading at the pile head, on the pile shaft, and at the pile base are denoted by T, T_s, and T_b, respectively. Vertical and rotational displacement at the pile head are denoted w and $\theta D/2$, respectively, where θ is angle of rotation and D is pile diameter. Rotary jacking is characterized by the pitch, p, of the pile motion, which is defined as the ratio of horizontal and vertical speeds evaluated at the outer surface of the pile (noting that this definition is inverted compared to the definition of the pitch of a screw thread).

Rotary-jacking installation occurs in a number of jacking strokes, similar to an axially jacked pile, although the machine can provide continuous rotation. The stroke length is limited by the length of the hydraulic rams used to apply axial force. At the end of a vertical jacking stroke rotation ceases while the chuck of the piling machine declutches from the pile and is returned to its initial position.

Figure 5.30 illustrates an idealized displacement and loading history of a closed-ended rotary-jacked pile into an idealized ground profile of linearly increasing resistance with depth. It is assumed that the pile is installed at a constant pitch (stage a-b) in four equal-length jacking strokes. At the end of each jacking stroke, the pile chuck declutches from the pile (stage b-c) and the axial and torque loads at the pile head are

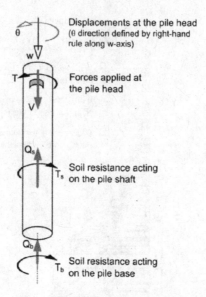

Figure 5.29 Resistances around the pile during rotary press-in operation (Deeks, 2008).

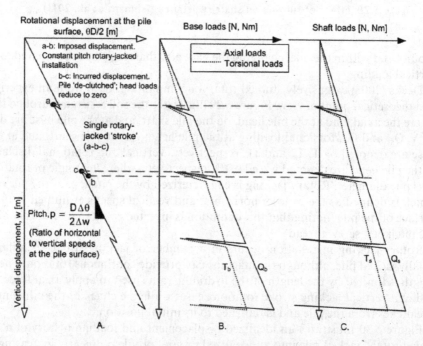

Figure 5.30 An idealized displacement and loading history of a closed-ended rotary-jacked pile (Deeks, 2008). (a) Change in rotational displacement with depth, (b) change in base load with depth, and (c) change in shaft load with depth.

reduced to zero. This is likely to result in compression and tensile residual loads on the base and shaft, respectively.

Deeks (2008) conducted a series of centrifuge tests using a closed-ended tubular pile installed by rotary jacking in a dense dry sand, in which the torque and axial loads were measured at the pile head and base. Thus, each component can be decomposed as shown in Figure 5.31.

Figure 5.32 gives the displacement and load histories with depth of two rotary-jacked installations (without and with rotary), with respect to base axial load, Q_b, and base torque, T_b. In test TW6015, the pile was installed with pure vertical movement (without

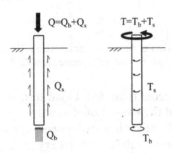

Figure 5.31 Decomposition of axial load and torque (Ishihara et al., 2015). (a) Components of axial load and (b) components of torque.

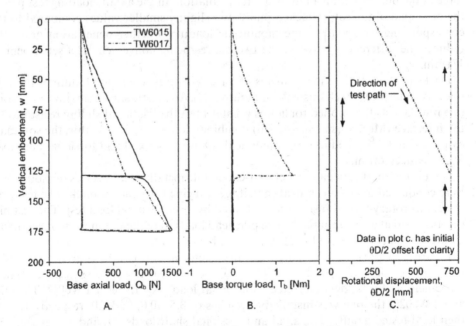

Figure 5.32 Effects of rotary jacking on base axial/torque load and rotational displacement histories during pile installation (Deeks, 2008). (a) Change in base axial load with vertical displacement, (b) change in base torque with vertical displacement, and (c) change in rotational displacement with vertical displacement.

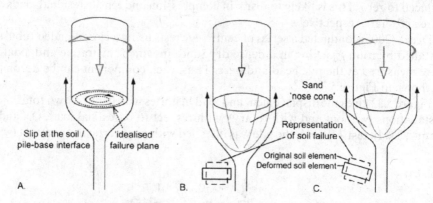

Figure 5.33 Proposed three failure mechanisms (Deeks, 2008). (a) Mechanism A, (b) mechanism B, and (c) mechanism C.

rotary) and was then subjected to an axial load test. In test TW6017, the pile was installed at a pitch of 3.5 and then load tested axially.

The loads Q_b and T_b of the both tests increase approximately linearly with depth during the installation phase for the case of rotary jacking. The base load of the rotary jacked pile is consistently smaller, by about 20%, than the base load of the monotonic installation. This demonstrates that to install a rotary-jacked pile requires less axial force at the pile base than a purely axial installation. In the axially loading test phase, the base load of the rotary-jacked pile shows slightly smaller values compared to the corresponding base load of the monotonic loading. Further penetration gradually reduces the differences between the two curves being convergent at a settlement of 175 mm.

Deeks proposed three mechanisms of base failure: (i) frictional failure at the soil/pile base interface (Mechanism A, in Figure 5.33), (ii) plastic slip occurring at the soil-pile interface as the applied torsion is greater than the frictional sliding resistance of the interface (Mechanism B), and (iii) combined loading failure within the soil mass (Mechanism C). He deduced the analytical solutions of axial and torsional base load for each mechanism.

Next is the shaft capacity of a closed-ended pile installation. A series of model tests were conducted in a geotechnical centrifuge, varying the value of pitch. The test program was rotary-jacked installation followed by a purely axial load test. The test pile was installed at a constant pitch. The pile head loads were then reduced to zero, before a purely axial compressive load test was conducted.

Figure 5.34 presents the shaft load-displacement histories of four rotary-jacked installations for four tests. In test TW6015, the pile was installed with pure vertical movement and was then subjected to an axial load test. In tests TW6017, TW6018, and TW6022, the pile was installed at pitches of 3.5, 10.0, and 1.0, respectively, and then load-tested axially. The axial and torsional shaft loads, Q_s and T_s, respectively increase nonlinearly with depth. The rates of increase $\Delta Q_s / \Delta w$ and $\Delta T_s / \Delta w$ increase with increasing depth, w. During the installation stage, except for test TW6022, the shaft axial loads of rotary-jacked installation piles are marginally smaller than that of the axially jacked installation pile. During the axial load test stage, there are notable

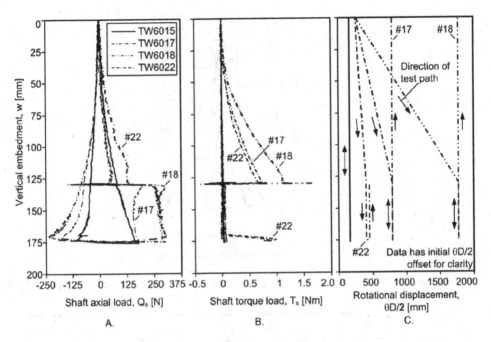

Figure 5.34 Shaft load-displacement histories of rotary-jacked installations (Deeks, 2008). (a) Change in base axial load with vertical displacement, (b) change in base torque with vertical displacement, and (c) change in rotational displacement with vertical displacement.

differences in the shaft axial load–depth relationship. The values of Q_s mobilized during the load test of the rotary-jacked installations are more than twice that mobilized during purely axial installation. The mean unit shaft resistance $\tau_{s, mean}$ is calculated as

$$\tau_{s, mean} = \left(\frac{Q_s}{|Q_s|}\right)\frac{1}{(\pi DL)}\left((Q_s)^2 + \left(\frac{2T_s}{D}\right)^2\right)^{0.5} \tag{5.14}$$

where L: the current depth of embedment.

The multiplier $(Q_s/|Q_s|)$ allows cases of net axially compressive $(Q_s/|Q_s|=1)$ and axial tensile $(Q_s/|Q_s|=-1)$ loading to be easily distinguished. Figure 5.35 plots the mean shaft shear stress at failure against embedded depth.

The purely vertical installation (TW6015) mobilizes a shaft capacity of 21 kPa at the end of installation. The rotary-jacked installations (TW6017, 6018, 6022) mobilize capacities of 43, 63, and 53 kPa, respectively, two to three times greater than the purely vertical installation.

The subsequent axial load test shows this increase in shaft capacity to be brittle in nature. In test TW6017, the shaft capacity reduces steadily to that measured in the purely axial installation and load test of TW6015. The axial load tests of TW6018 and TW6022 also show reductions in capacity with displacement, though in those cases the mean unit shaft resistance is till twice that measured in the purely axial installation.

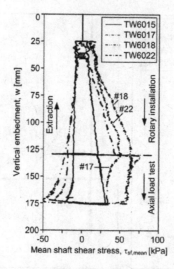

Figure 5.35 Mean shaft shear stress at failure against embedded depth (Deeks, 2008).

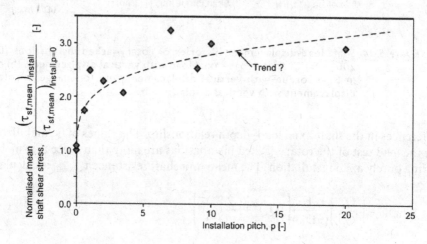

Figure 5.36 Mean shaft capacity increases with the pitch of installation (Deeks, 2008).

This increased capacity is confirmed in the subsequent tensile load test. The increase in mean shaft capacity with increasing pitch of installation is illustrated in Figure 5.36, showing that the mean shaft capacity increases with the pitch of installation.

5.4 Performance of a single pile installed in press-in piling

5.4.1 Introduction

Preceding sections mainly focus on the response of a pile during installation. This section describes the subsequent performance of a single pile installed by the press-in piling under various loading conditions, including vertical and horizontal monotonic loading and cyclic loading. This section also presents some results of numerical analysis, demonstrating the effect of installation on the performance of installed pile.

There exists usually in practice a time gap between the installation stage and the loading stage. Pile capacity often increases with the elapse of time. These effects are called 'setup'. Since there are a number of research outcomes available that are related to setup, this chapter does not discuss the matter in detail.

5.4.2 Residual stresses in piles and surrounding soils

After the installation of the pile is completed, the force at the pile head reduces to zero. However, there are residual stresses locked in the pile. This is because the surrounding soil prevents the pile from rebounding upward, producing negative friction resistance at the upper part of the shaft and positive friction resistance at the lower part of the shaft. At the same time, a portion of the base resistance still remains at the pile tip. The residual load distribution in the pile is schematically illustrated in Figure 5.37.

The experimental observation (White & Bolton, 2004) suggested that the penetration mechanism is more comparable to cavity expansion, with the displacement vectors radiating from the pile tip downward and outward. A numerical simulation illustrates the stresses in the ground after installation, in terms of the difference between bored piles and jacked-in piles, as shown in Figure 5.38.

In the numerical analysis, for the modeling of the stress condition after jacked-in installation, a solid pile with half its real diameter is placed in soil first and then expanded to the full pile diameter to achieve the radial stress of the soil surrounding a jacked pile.

5.4.3 Stiffness and capacity

As was described in the preceding sections, the centrifuge study of the effect of surging revealed that base resistance may increase depending on the amount of stroke and

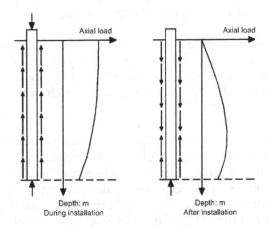

Figure 5.37 Residual load distribution in the pile (Yetginer et al., 2006). (a) Axial load distribution during installation and (b) axial load distribution after installation.

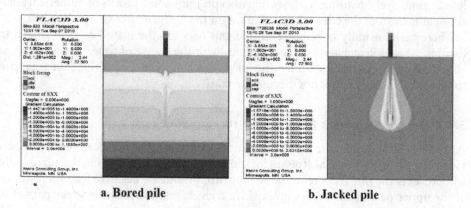

a. Bored pile   b. Jacked pile

Figure 5.38 The radial soil stress before lateral loading (Li, 2010). (a) Bored pile and (b) jacked pile.

on whether the soil is dry or saturated. Furthermore, the shaft resistance appreciably reduces. Field test data generally confirmed the experimental observation.

The centrifuge study of the effect of rotary-jacked installation showed that base stiffness is almost comparable to those of monotonic installation and shaft resistance increases almost by double.

Field test data presented by Deeks et al. (2005) shows that the stiffness of jacked piles is considerably higher than conventional driven or bored piles. Existing guidance for the design of bored and driven piles is collated in Figure 5.39 and compared with jacked pile results from this investigation and existing published data. The characteristic secant base stiffness of the jacked piles at a settlement of 2% D is more than two or ten times greater than recommended design values for driven and bored piles, respectively.

Li (2010) conducted a series of centrifuge tests, looking into the effects of the cyclic loading of piles on the load–settlement relationship. Figure 5.40 shows centrifuge tests of monotonically jacked pile subject to ten cycles of axially cyclic loading. It is seen from the figure that base load and secant stiffness decrease with an increasing number of cycles.

It may be of some use for the readers to introduce a few results of numerical analysis. Li (2010) analyzes the response of bored and jacked pile subject to lateral loading using a numerical modeling program called "Fast Lagrangian Analysis of Continua in Three Dimensions" (FLAC3D). The numerical result of the jacked pile subject to lateral loading is described.

Pile heads are fixed without any rotation and moved laterally after installation. The lateral displacement of the pile head is approximately 0.15 m, and the deflection of the pile shafts is as presented in Figure 5.41. The magnitude of deformation is amplified here in order to obtain a better observation. The deflection of bored and jacked piles is evident at shallow depth, and it decreases rapidly with increasing depth in the soil. The bored pile shaft rotates to some extent, and the depth of the rotation center is about 7.5 m (75% of the embedment depth), which is consistent with that found by Broms (1964). On the other hand, the jacked pile deformed in the long pile mode, and the pile shaft has negligible deflection below a depth of ~6.0 m.

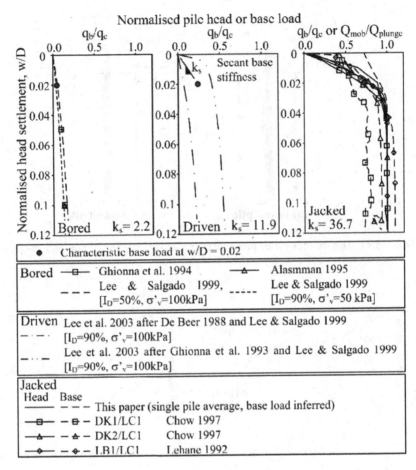

Figure 5.39 Relative stiffness of jacked, driven, and bored piles (Deeks et al., 2005).

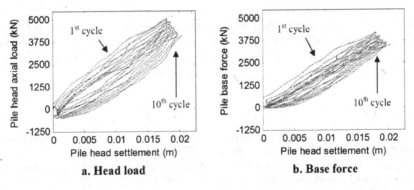

Figure 5.40 Monotonic-jacked head axial force–displacement curves, in displacement-control cycles (Li, 2010). (a) Head load and (b) base force.

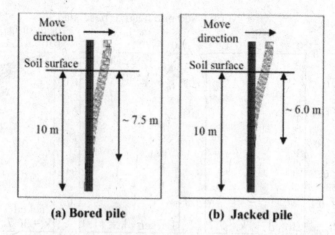

(a) Bored pile   **(b) Jacked pile**

Figure 5.41 The deflection of pile shafts (Li, 2010). (a) Bored pile and (b) jacked pile.

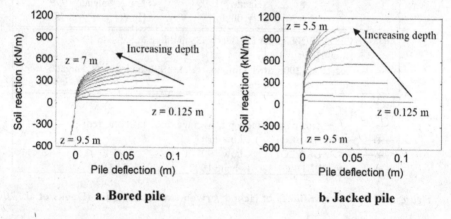

a. Bored pile   **b. Jacked pile**

Figure 5.42 p–y curves obtained in a numerical simulation (Li, 2010). (a) Bored pile and (b) jacked pile.

In FLAC3D simulations, the relationship between the soil reaction per unit length of pile and the pile deflection, i.e. p–y curves, was obtained at depths with 0.5 m interval, as shown in Figure 5.42. It is found that p–y curves of the jacked pile are much stiffer than those of the bored pile at corresponding depths.

References

Aabøe, R. (2001) Environmental effects related to the construction of a cut and cover road tunnel, *Nordic Road & Transport Research*, Vol. 13, Issue 1, pp. 4–5.

Broms, B.B. (1964) The lateral resistance of piles in cohesionless soils, *Journal of the Soil Mechanics and Foundations Division, ASCE*, Vol. 90, Issue 3, pp. 123–156.

BS5228 (1992) *Noise Control on Construction and Open Sites - Part 4: Code of Practice for Noise and Vibration Control Applicable to Piling Operations*. London, British Standards Institution.

Burali d'Arezzo, F. (2015) Installation effects due to pile surging in sand, PhD thesis, Cambridge University.

Deeks, A.D. (2008) An investigation into the strength and stiffness of jacking pile in sand, PhD thesis, Cambridge University.

Deeks, A.D., White, D.J. and Bolton, M.D. (2005) A comparison of jacked, driven and bored piles in sand, *Proceedings of the 16th International Conference on Soil Mechanics and Geotechnical Engineering*, pp. 2103–2106.

Eurocode 3 (1992) *Eurocode 3. Design of Steel Structures – Part 5: Piling. BS EN 1993–5:1992 (draft)*. London, British Standards Institution.

Head, J.M. and Jardine, F.M. (1992) Ground-borne vibrations arising from piling, Construction Industry Research and Information Association (CIRIA), UK. Technical Note 142.

IPA (2016) *Press-in Retaining Structures: A Handbook*. International Press-in Association.

Ishihara, Y., Haigh, S. and Bolton, M.D. (2015) Estimating base resistance and N value in rotary press-in, *Soils and Foundation*, Vol. 55, Issue 4, pp. 788–797.

Ishihara, Y., Okada, K., Nishigawa, M., Ogawa, N., Horikawa, Y. and Kitamura, A. (2011) Estimating PPT data Via CPT-based design method, *Proceedings of 3rd IPA International Workshop in Shanghai*, pp. 84–94.

Li, Z. (2010) Piled foundations subjected to cyclic loads or earthquakes, PhD thesis, University of Cambridge.

Li, K. S., Ho, N. C. L., Lee, P. K. K., Tham, L. G. & Lam, J. (2003) Chapter 2: jacked piles. In Li, K. S. (Ed.) *Design and Construction of Driven and Jacked Piles*, Hong Kong, Centre for Research and Professional Development.

Rockhill, D.J., Bolton, M.D. and White, D.J. (2003) Ground-borne vibrations due to press-in piling operations. In *BGA International Conference on Foundations, Innovations, Observations, Design and Practice*, Thomas Telford Ltd, pp. 743–756.

White, D.J. (2002) An investigation into the behavior of pressed-in piles, PhD thesis, Cambridge University.

White, D.J. and Bolton, M.D. (2004) Displacement and strain paths during plane-strain model pile installation in sand, *Geotechnique*, Vol. 54, Issue 6, pp. 375–397.

White, D.J. and Deeks, A.D. (2007) Recent research into the behaviour of jacked foundation piles, In *Advances in Deep Foundations: International Workshop on Recent Advances of Deep Foundations*, pp. 3–26.

White, D., Finlay, T., Bolton, M. and Bearss, G. (2002) Press-in piling: Ground vibration and noise during pile installation, *Proceedings of the International Deep Foundations Congress*, Orlando, USA, pp. 363–371.

White, D.J., Sidhu, H.K., Finlay, T.C.R., Bolton, M.D. and Nagayama, T. (2000) Press-in piling: The influence of plugging on driveability, *8th International Conference of the Deep Foundations Institute*, New York, pp. 299–310.

Yetginer, A.G., White, D.J. and Bolton, M.D. (2006) Field measurements of the stiffness of jacked piles and pile groups, *Geotechnique*, Vol. 56, Issue 5, pp. 349–354.

Appendix

List of related publications

For the convenience and benefit of the readers, publications as many as about 200 related to press-in technology have been complied. Table A1 provides a classification of research topics. The papers are classified in terms of research topics and plotted against the year of publication from the 1960s to the 2010s in Tables A2a and 2b.

Publication details are listed below in the order of classification category in the table.

Table AI List of research topics

Research topics to understand and improve Press-in Piling technologies

- Automatic acquisition and application of press-in data
- Reduction of the resistance of the ground during press-in
  - Effect of penetration rate
  - Effect of repetitive upward and downward motion (surging)
  - Effect of water jetting at pile toe
  - Effect of Rotary press-in piling
- Time dependent increase, set-up, of pile capacity after pile installation
- Effect of dilatancy or suction to increase the pull-out resistance
- Performance of pressed-in (jacked) piles
  - Vertical bearing capacity and stiffness
  - Effect of cyclic jacking at a constant depth to increase vertical bearing capacity
  - Vertical bearing characteristics of pressed-in sheet piles
  - Horizontal bearing characteristics of pressed-in piles
- Group effect on vertical bearing capacity

Research topics to understand the function and the performance of Embedded Walls and Structures

- Performance of steel sheet pile walls
- Performance of embedded wall against surcharge loading on retained soil
- Embedded wall to reduce ground deformation associated with adjacent construction or increased embankment load
- Performance of pile walls embedded in hard ground
- Performance of embedded wall enclosure as countermeasures against liquefaction
- Performance of structures with piles as countermeasures against tsunami
- Researches on levees reinforced by double sheet pile walls
- Others

Table A2-a Classification of papers according to the research topics 1/2

		~1999	2000s	2010s
Research topics to understand and improve press-in piling technologies	Reduction of the resistance of the ground during press-in			
	Automatic acquisition and application of press-in data	[1]	[2] [3] / [4]	[5] [6^j] / [7] [8] [9] [10] [11] / [12] [13] [14] / [15^j] [16] [17^j] [18^j] [19]
	Effect of penetration rate	[20^j]	[21] / [22]	[23] [24] [25] / [26] [27] [28^j] [29] [30]
	Effect of repetitive upward and downward motion (surging)	[31]	[32] / [33]	[34] [35] / [36] [8] [37] [38] / [39] [40]
	Effect of water jetting at pile toe	[41]		[30] [42] / [39] [40] [43] [44] [45]
	Effect of rotary press-in piling			[27] [8] [29] [46] [47^j] [48] / [45] [49]

(Continued)

Table A2-a Classification of papers according to the research topics 1/2

		~1999	2000s	2010s
Research topics to understand and improve press-in piling technologies	Time-dependent increase, set-up, of pile capacity after pile installation		[50] [51]	[52] [35] [54] [53] [30] [55]
	Effect of dilatancy or suction to increase the pullout resistance			[56i] [57]
Performance of pressed-in (jacked) piles	Vertical bearing capacity and stiffness	[58i]	[59] [60] [61] [62] [63] [64] [66] [68] [72] [65] [67] [70] [71] [69]	[73] [76] [34] [79] [80] [74] [35] [77] [81] [75] [78] [27] [82] [83] [84] [85] [86]
	Effect of cyclic jacking at a constant depth to increase vertical bearing capacity			[87] [84] [88]
	Vertical bearing characteristics of pressed-in sheet piles			[89] [90] [91]
	Horizontal bearing characteristics of pressed-in piles			[92] [93]
	Group effect on vertical bearing capacity		[67] [72] [94] [95]	

Note: Papers with the suffix J (e.g., 20J) are written in Japanese.
Reference number of a paper, dealing with more than one topic, will appear under the different topics.

Table A2-b Classification of papers according to the research topics 2/2

		~1999	2000s	2010s			
Research topics to understand the function and the performance of embedded walls and structures	Performance of steel sheet pile walls			[96ʲ] [97ʲ]	[98ʲ] [99ʲ] [100ʲ]	[101ʲ]	[102] [103] [104] [105] [106] [107] [108] [109] [110] [111] [112]
	Performance of embedded wall against surcharge loading on retained soil				[113ʲ] [114ʲ]	[115ʲ] [116ʲ] [117ʲ]	[118] [119ʲ]
	Embedded wall to reduce ground deformation associated with adjacent construction or increased embankment load			[120ʲ]			[121] [122] [123ʲ] [124] [125] [126] [127]
	Performance of pile walls embedded in hard ground						[122] [128] [129] [130] [131] [132]

(Continued)

Table A2-b Classification of papers according to the research topics 1/2

Research topics to understand the function and the performance of embedded walls and structures	2000s	2010s
Performance of embedded wall enclosure as countermeasures against liquefaction	[133j] [134j] [135j] [136j]	[137j] [138j] [139j] [140j] [141]
Performance of structures with piles as countermeasures against tsunami		[142j] [143j] [144j] [145j] [146j] [147j] [148j] [149j] [150j] [151] [152] [153j] [154j] [155] [156] [157] [158]
Researches on levees reinforced by double sheet pile walls	[159j] [160j] [161j] [162j] [163j]	[164j] [165j] [166j] [167j] [168j] [169j] [170j] [171j] [172j] [173j] [174j] [175j]
Others	[176] [177] [178j] [179] [180] [181] [182j] [183] [184] [185j]	[187] [188] [189] [190] [191] [192] [193] [194] [195] [108] [196]

Note: Papers with the suffix J (e.g., 20ᴶ) are written in Japanese.
Reference number of a paper, dealing with more than one topic, will appear under the different topics.

Research topics to understand and improve press-in piling technologies

Automatic acquisition and application of press-in data

1) Teale, R. (1965): The concept of specific energy in rock drilling, International Journal of Rock Mechanics & Mining Sciences, Vol. 2, pp. 57–73.

2) Robertson, P.K. (1990): Soil classification using the cone penetration test. Canadian Geotechnical Journal, 27, pp. 151–158.

3) Jefferies, M.G. and Davies, M.P. (1993): Use of CPTu to estimate equivalent SPT N60, Geotechnical Testing Journal, GTJODJ, Vol. 16, No. 4, pp. 458–468.

4) Finlay, T.C.R., White, D.J. and Bolton, M.D. (2001): Press-in piling: the influence of instrumented steel tubular piles with and without driving shoes, Proc. of 5th International Conference on Deep Foundation Practice, Singapore, pp. 199–208.

5) Ishihara, Y., Ogawa, N., Horiwakawa, Y., Kinoshita, S., Nagayama, T., Kitamura, A. and Tagaya, K. (2009): Utilization of pile penetration test data for ground information, Proc. of 2nd IPA International Workshop in New Orleans, Press-in Engineering 2009, pp. 105–120.

6) Ishihara, Y., Ogawa, N., Kinoshita, S. and Tagaya, K. (2009): Study on soil classification and equivalent N value based on PPT data, Proc. of Recent Sounding Technology and Ground Evaluation Symposium, pp. 85–90. (in Japanese).

7) Ishihara, Y., Nagayama, T. and Tagaya, K. (2010): Interpretation PPT data for more efficient pile construction, Proc. of the 11th International Conference on Geotechnical Challenges in Urban Regeneration, London, UK, CD, 8p.

8) Ishihara, Y., Okada, K., Nishigawa, M., Ogawa, N., Horikawa, Y. and Kitamura, A. (2011): Estimating PPT data via CPT-based design method, Proc. of 3rd IPA International Workshop in Shanghai, Press-in Engineering 2011, pp. 84–94.

9) Ogawa, N., Nishigawa, M. and Ishihara, Y. (2012): Estimation of soil type and N-value from data in press-in piling construction. Testing and Design Methods for Deep Foundations, IS-Kanazawa 2012, pp. 597–604.

10) Ishihara, Y., Ogawa, N., Lei, M., Okada, K., Nishigawa, M. and Kitamura, A. (2013): Estimation of N value and soil type from PPT data in standard press-in and press-in with Augering, Proc. of 4th IPA International Workshop in Singapore, Press-in Engineering 2013, pp. 116–129.

11) Ishihara, Y., Ogawa, N. and Lei, M. (2014): Estimating N value from data in press-in piling, Proc. of International Conference on Piling & Deep Foundations, Stockholm, Sweden, USB, 10p.

12) Ishihara, Y., Haigh, S.K. and Bolton, M.D. (2015): Estimating base resistance and N values in rotary press-in, Soils and Foundations, Vol. 55, No. 4, pp. 788–797.

13) Ishihara, Y., Ogawa, N., Okada, K. and Kitamura, A. (2015): Estimating subsurface information from data in press-in piling, Proc. of International Workshop, Press-in Engineering 2015, pp. 53–67.

14) Ishihara, Y., Haigh, S.K. and Bolton, M.D. (2015): Estimating base resistance and N value in rotary press-in, Soils and Foundations, Vol. 55, No. 4, pp. 788–797.

15) International Press-in Association (IPA). (2017): Technical Material on the Use of Piling Data in the Press-in Method, I. Estimation of Subsurface Information, 63p. (in Japanese).

16) Ishihara, Y. (2018): Use of press-in piling data for automatic operation of press-in machines and estimation of subsurface information, Proc. of the First International Conference on Press-in Engineering 2018, Kochi, pp. 651–660.

17) Ishihara, Y., Kikuchi, Y. and Koseki, J. (2018): Model test on correlation between plugging condition and incremental filling ratio during press-in of open-ended tubular pile, Proc. of the 53rd Japan National Conference on Geotechnical Engineering, pp. 1321–1322. (in Japanese).

18) Ishihara, Y., Nose, T., Hamada, K and Matsuoka, T. (2018): Automatic operation using piling data in the press-in method, The Japanese Geotechnical Society, Vol. 66, No. 1, pp. 26–27. (in Japanese).

19) Furuichi, H., Nakayama, T. and Suzuki, N. (2019): Countermeasure against storm surge, and estimation of soil type and N value by standard press-in method of steel sheet pile, Press-in Piling Case History, Volume 1, pp. 151–152.

Reduction of the resistance of the ground during press-in

Effect of penetration rate

20) Yamahara, H. (1964): Plugging effect and bearing mechanism of steel pipe pile (Part-1), Transactions of the Architectural Institute of Japan, Vol. 96, pp. 28–35. (in Japanese).

21) Randolph, M.F., Leong, E.C. and Houlsby, G.T. (1991): One-dimensional analysis of soil plugs in pipe piles, Geotechnique, Vol. 41, No. 4, pp. 587–598.

22) Randolph, M. F. and Hope, S. (2004): Effect of cone velocity on cone resistance and excess pore pressures, Proc. of the IS Osaka – Engineering Practice and Performance of Soft Deposits, pp. 147–152.

23) Silva, M.F., White, D.J. and Bolton, M.D. (2006): An analytical study of the effect of penetration rate on piezocone tests in clay, International Journal for Numerical and Analytical Methods in Geomechanics, Vol. 30, pp. 501–527.

24) Jackson, A.M., White, D.J., Bolton, M.D. and Nagayama, T. (2008): Pore pressure effects in sand and silt during pile jacking, Proc. of the 2nd BGA International Conference on Foundations, CD, pp. 575–586.

25) Shepley, P. (2009): An investigation into the plugging of open-ended jacked-in tubular piles, M. Eng. thesis, University of Cambridge, 47p.

26) Jeager, R.A., DeJong, J.T., Boulanger, R.W., Low, H.E. and Randolph, M.F. (2010): Variable penetration rate CPT in an intermediate soil, 2nd International Symposium on Cone Penetration Testing, CPT 2010, 8p.

27) White, D.J., Deeks, A.D. and Ishihara, Y. (2010): Novel piling: axial and rotary jacking, Proc. of the 11th International Conference on Geotechnical Challenges in Urban Regeneration, London, UK, CD, 24p.

28) Watanabe, K. and Sahara, M. (2012): Effect of Loading Rate on Bearing Capacity and Soil Spring of Pile Foundations, Report of Obayashi Technical Research Institute, No. 76, pp. 1–8. (in Japanese).

29) Winstanley, T. (2012): The significance of pore water pressures on press-in piles. M.Eng. Project Report, Cambridge University Department of Engineering, 50p.
30) Bolton, M.D., Haigh, S.K., Shepley, P. and White, D.J. and Deeks, A.D. (2013): Identifying ground interaction mechanisms for press-in piles, Proc. of 4th IPA International Workshop in Singapore, Press-in Engineering 2013, pp. 84–95.

Effect of repetitive upward and downward motion (surging)

31) Herrema, E. P. (1980): Predicting pile drivability: heather as an illustration of the "friction fatigue" theory, Ground Engineering, pp. 15–20.
32) White, D.J. and Bolton, M.D. (2002): Observing friction fatigue on a jacked pile, Proc. of the Workshop on Constitutive and Centrifuge Modelling: Two extremes, ed. Springman, S.M., Balkema, pp. 347–354.
33) Lehane, B.M., Schneider, J.A. and Xu, X. (2005): The UWA-05 method for prediction of axial capacity of driven piles in sand, International Symposium on Frontiers in Offshore Geotechnics, pp. 683–689.
34) Lehane, B.M., Schneider, J.A. and Xu, X. (2007): CPT-based design of displacement piles in siliceous sands, Advances in Deep Foundations, pp. 69–86.
35) White, D.J. and Deeks, A.D. (2007): Recent research into the behaviour of jacked foundation piles, Advances in Deep Foundations, pp. 3–26.
36) Delano, O. (2010): The application of surging on jacked-in piles, M.Eng. Project Report, Cambridge University Department of Engineering, 49p.
8) Ishihara, Y., Okada, K., Nishigawa, M., Ogawa, N., Horikawa, Y. and Kitamura, A. (2011): Estimating PPT data via CPT-based design method, Proc. of 3rd IPA International Workshop in Shanghai, Press-in Engineering 2011, pp. 84–94.
37) Ogawa, N., Ishihara, Y., Nishigawa, M. and Kitamura, A. (2011): Effect of surging in press-in piling: shaft resistance and pore water pressure. Proc. of the 3rd IPA International Workshop in Shanghai, Press-in Engineering 2011, pp. 101–106.
38) Burali d'Arezzo, F., Haigh, S.K. and Ishihara, Y. (2014): Modelling of jacked piles in centrifuge, Proc. of the 8th International Conference on Physical Modelling in Geotechnics, pp. 807–812.
39) Burali d'Arezzo, F. (2015): Installation effects due to pile surging in sand, PhD thesis, Cambridge University.
40) Burali d'Arezzo, F., Haigh, S.K., Talesnick, M and Ishihara, Y. (2015): Measuring horizontal stresses during jacked pile installation, Proc. of the Institution of Civil Engineers, Vol. 168, Issue 4, pp. 306–318.

Effect of water jetting at pile toe

41) Tsinker, G.P. (1988): Pile jetting, Journal of Geotechnical Engineering, Vol. 114, No. 3, pp. 326–334.
30) Bolton, M.D., Haigh, S.K., Shepley, P. and White, D.J. and Deeks, A.D. (2013): Identifying ground interaction mechanisms for press-in piles, Proc. of 4th IPA International Workshop in Singapore, Press-in Engineering 2013, pp. 84–95.
42) Shepley, P. (2013): Water injection to assist pile jacking, PhD thesis, Cambridge University, 235p.

43) Gillow, M. (2018): Water jetting for sheet piling in sandy soils. M.Eng. Project Report, Cambridge University Department of Engineering, 49p.

44) Gillow, M., Haigh, S.K., Ishihara, Y., Ogawa, N. and Okada, K. (2018): Water jetting for sheet piling, Proc. of the First International Conference on Press-in Engineering 2018, Kochi, pp. 335–342.

45) Ishihara, Y. and Haigh, S.K. (2018): Cambridge-Giken collaborative working on pile-soil interaction mechanisms, Proc. of the First International Conference on Press-in Engineering 2018, Kochi, pp. 23–45.

Effect of Rotary press-in piling

27) White, D.J., Deeks, A.D. and Ishihara, Y. (2010): Novel piling: axial and rotary jacking, Proc. of the 11th International Conference on Geotechnical Challenges in Urban Regeneration, London, UK, CD, 24p.

8) Ishihara, Y., Okada, K., Nishigawa, M., Ogawa, N., Horikawa, Y. and Kitamura, A. (2011): Estimating PPT data via CPT-based design method, Proc. of 3rd IPA International Workshop in Shanghai, Press-in Engineering 2011, pp. 84–94.

46) Bond, T. (2011): Rotary jacking of tubular piles, M.Eng. Project Report, Cambridge University Department of Engineering, 50p.

47) Nishigawa, M., Okada, K., Bond, T., Yamane, T., Ishihara, Y. and Kitamura, A. (2011): Reduction of friction in rotary jacking, Proc. of International Workshop, Press-in Engineering 2011, pp. 84–94.

29) Winstanley, T. (2012): The significance of pore water pressures on press-in piles. M.Eng. Project Report, Cambridge University Department of Engineering, 50p.

48) Hazla, E. (2013): Rotary press-in piling in hard ground, M.Eng. Project Report, Cambridge University Department of Engineering, 50p.

45) Ishihara, Y. and Haigh, S.K. (2018): Cambridge-Giken collaborative working on pile-soil interaction mechanisms, Proc. of the First International Conference on Press-in Engineering 2018, Kochi, pp. 23–45.

49) Frick, D., Schmoor, K.A., Gütz, P. and Achmus, M. (2018): Model testing of rotary jacking open ended tubular piles in saturated non-cohesive soil, Physical Modelling in Geotechnics, London, pp. 1347–1352.

Time dependent increase, set-up, of pile capacity after pile installation

50) Zhao, Y. (2002): Pile set-up in sand. M.Eng. Project Report, Cambridge University Department of Engineering, 48p.

51) Komurka, V.E., Wagner, A.B. and Edil, T.B. (2003): Estimating soil/pile set-up, Wisconsin Highway Research Program #0092-00-14, Final Report, 43p.

52) Zhao, Y. and White, D.J. (2006): A model-scale investigation into 'set-up' of displacement piles in sand, Physical Modelling in Geotechnics -6th ICPMG, pp. 889–894.

35) White, D.J. and Deeks, A.D. (2007): Recent research into the behaviour of jacked foundation piles, Advances in Deep Foundations, pp. 3–26.

53) Jackson, A. (2007): The setup of jacked piles, M.Eng. Project Report, Cambridge University Department of Engineering, 49p.

54) Zhang, L.M., and Choi, S.Y. (2008). Time-dependent behaviour of jacked piles, Proc. of 2nd IPA International Workshop in New Orleans, Press-in Engineering 2009, pp. 89–95.

30) Bolton, M.D., Haigh, S.K., Shepley, P. and White, D.J. and Deeks, A.D. (2013): Identifying ground interaction mechanisms for press-in piles, Proc. of 4th IPA International Workshop in Singapore, Press-in Engineering 2013, pp. 84–95.

55) Nozaki, T. (2018): Recovery of skin friction of Cambridge Gault clay with time effect, Proc. of the First International Conference on Press-in Engineering 2018, Kochi, pp. 497–500.

Effect of dilatancy or suction to increase the pullout resistance

56) Kato, T. and Kokusho, T. (2012): Rate-dependent pull-out bearing capacity of piles by similitude model tests using seepage force, Journal of Japan Society of Civil Engineers, Ser. C (Geotechnical Engineering), Vol. 68, No. 1, pp. 117–126. (in Japanese).

57) Stevens, G. (2015): Mechanism of water binding during press-in in sand. M.Eng. Project Report, Cambridge University Department of Engineering, 50p.

Performance of pressed-in (jacked) piles

Vertical bearing capacity and stiffness

58) Fujita, K. and Ueda, K. (1971): On the elapsed time after the completion of pile driving and its capacity, Proc. of the Symposium on Problems in the Method of Vertical Load Test on Piles, Soils and Foundations, Vol. 19, No. 6, p. 28. (in Japanese).

59) Det Norske Veritas. (DNV) (1992): Foundations, Classification Notes, No. 30.4, 54p.

60) Ghionna, V.N., Jamiolkowski, M. and Lancellotta, R. (1993): Base capacity of bored piles in sands from in situ tests. Proc. of the International Conference on Bored and Augured Piles II, Balkema, pp. 67–75.

61) Finnie, I. M.S. and Randolph, M.F. (1994): Punch-through and liquefaction-induced failure of shallow foundations on calcareous sediments, Proc. of International Conference on Behaviour of Offshore Structures, BOSS'94, pp. 217–230.

62) White, D. J. (1998): Deep penetration in sand. M.Eng. Project Report, Cambridge University Department of Engineering, 51p.

63) White, D.J., Sidhu, H.K., Finlay, T.C.R., and Bolton, M.D. (2000): Press-in Piling: The influence of plugging on drivability, Proc. of 8th International Conference of the Deep Foundations Institute, New York, pp. 299–310.

64) Lehane, B.M. and Gavin, K.G. (2001): Base resistance of jacked pipe piles in sand, ASCE Journal of Geotechnical and Geoenvironmental Engineering, Vol. 127, No. 6, pp. 473–480.

65) Liyanapathirana, D.S., Deeks, A.D. and Randolph, M.F. (2001): Numerical modelling of the driving response of thin-walled open-ended piles, International Journal for Numerical and Analytical Methods in Geomechanics, Vol. 25, No. 9, pp. 933–953.

66) White, D.J. (2002): An investigation into the behavior of pressed-in piles, PhD thesis, Cambridge University.

67) White, D.J. (2002): The use of pressed-in H-piles for larger foundation structures, Proc. of the 7th BGA Young Geotechnical Engineers' Symposium, Vol. 1, 2p.

68) White, D.J. (2003): An urban foundation system using jacked H-piles, Proc. of 2nd International Young Geotechnical Engineers' Conference, Vol. 1, pp. 49–50.

69) White, D.J., Bolton, M.D. and Wako, C. (2003): A novel urban foundation system using pressed-in H piles, Proc. of 13th European Conference on Soil Mechanics and Geotechnical Engineering, Vol. 2, pp. 425–430.

70) Yetginer, A.G. (2003): Press-in piling. M.Eng. Project Report, Cambridge University Department of Engineering, 50p.

71) Yetginer, A.G., White, D.J. and Bolton, M.D. (2003): Press-in piling: field testing of cell foundations, Proc. of the International Conference organised by British Geotechnical Association, Vol. 1, pp. 963–974.

72) Deeks, A.D. (2004): An investigation into the strength and stiffness of jacked, driven and bored piles, M.Eng. Project Report, Cambridge University Department of Engineering, 51p.

73) Berardi, R. and Bovolenta, R. (2005): Pile-settlement evaluation using field stiffness non-linearity, Proc. of the Institution of Civil Engineers: Geotechnical Engineering, Vol. 158, Issue 1, pp. 35–44.

74) Lehane, B.M., Schneider, J.A. and Xu, X. (2005): CPT based design of driven piles in sand for offshore structures. Report, the University of Western Australia, GEO: 05345, 46p.

75) White, D.J. and Bolton, M.D. (2005): Comparing CPT and pile base resistance in sand. Proc. of the Institution of Civil Engineers, Geotechnical Engineering 158, pp. 3–14.

76) Dingle, H. (2006): The testing and analysis of jacked foundation piles, M.Eng. Project Report, Cambridge University Department of Engineering, 50p.

34) Lehane, B.M., Schneider, J.A. and Xu, X. (2007): CPT-based design of displacement piles in siliceous sands, Advances in Deep Foundations, pp. 69–86.

35) White, D.J. and Deeks, A.D. (2007): Recent research into the behaviour of jacked foundation piles, Advances in Deep Foundations, pp. 3–26.

77) Deeks, A.D. and White, D.J. (2007): Centrifuge modelling of the base response of closed-ended jacked piles, Advances in Deep Foundations, pp. 241–251.

78) Zhang, L.M., Ng, C.W.W., and Chan, F. (2007): Centrifuge modelling of construction of jacked pile in decomposed granite. Proc. 13th Asian Regional Conference on Soil Mechanics and Geotechnical Engineering, 10–14th December 2007, Kolkata-700045, India, In CD Rom.

79) Deeks, A.D. (2008): An investigation into the strength and stiffness of jacked piles in sand, PhD thesis, Cambridge University.

80) Ogawa, N., Ishihara, Y., Yokotobi, T., Kinohshita, S., Nagayama, T., Kitamura, A. and Tagaya, K. (2009): Soil plug behavior of open-ended tubular pile during press-in, Proc. of 2nd IPA International Workshop in New Orleans, pp. 121–129.

81) Zhang, L.M., and Wang, H. (2009). Field study of construction effects in jacked and driven steel H-piles. Geotechnique, Vol. 59, No. 1, pp. 63–69.

27) White, D.J., Deeks, A.D. and Ishihara, Y. (2010): Novel piling: axial and rotary jacking, Proc. of the 11th International Conference on Geotechnical Challenges in Urban Regeneration, London, UK, CD, 24p.

82) Li, Z., Bolton, M.D. and Haigh, S.K. (2010): The behavior of a single pile under cyclic axial loads, Proc. of the 2010 GeoShanghai International Conference on Deep Foundations and Geotechnical In Site Testing, ASCE Geotechnical Special Publications (205 GSP), pp. 143–148.

83) Okada, K. and Ishihara, Y. (2012): Estimating bearing capacity and jacking force for rotary jacking, Testing and Design Methods for Deep Foundations, IS-Kanazawa, pp. 605–614.

84) Burali d'Arezzo, F., Haigh, S. K. and Ishihara, Y. (2013): Cyclic jacking of piles in silt and sand, Installation Effects in Geotechnical Engineering, Proc. of the International Conference on Installation Effects in Geotechnical Engineering, ICIEGE 2013, pp. 86–91.

85) Ishihara, Y., Okada, K., Yokotobi, T. and Kitamura, A. (2016): Pull-out resistance of a large diameter steel tubular pile installed by rotary cutting press-in, Proc. of the Third International Conference Geotec Hanoi 2016- Geotechnics for Sustainable Infrastructure Development, pp. 141–147.

86) Watanabe, K., Kishi, H. and Ohashi, F. (2018): In-situ load test of press-in steel pipe pile for seismic isolation retrofit, Proc. of the First International Conference on Press-in Engineering 2018, Kochi, pp. 571–576.

Effect of cyclic jacking at a constant depth to increase vertical bearing capacity

87) Zhang, L.M., and Choi, S.Y. (2008): Cyclic loading effects for jacked piles, Proc. of 2nd IPA International Workshop in New Orleans, Press-in Engineering 2009, pp. 96–104.

84) Burali d'Arezzo, F., Haigh, S. K. and Ishihara, Y. (2013): Cyclic jacking of piles in silt and sand, Installation Effects in Geotechnical Engineering, Proc. of the International Conference on Installation Effects in Geotechnical Engineering, ICIEGE 2013, pp. 86–91.

88) Moriyasu, S., Meguro, H., Matsumoto, T., Kobayashi, S. and Shimono, S. (2016): Influence of surging and jack-in pile installation methods on pile performance observed in model load tests in dry sand grounds, Proc. of 19 Southeast Asian Geotechnical Society, Kuala Lumpur, Malaysia, pp. 621–626.

Vertical bearing characteristics of pressed-in sheet piles

89) Taenaka, S., Otani, J., Tatsuta, M. and Nishiumi, K. (2006): Vertical bearing capacity of steel sheet piles, Proc. of the Sixth International Conference on Physical Modelling in Geotechnics, Vol. 1, pp. 881–888.

90) Taenaka, S., White, D. J., Randolph, M. F., Nakayama, H. and Nishiumi, K. (2008): The shape effect of cross-sectional shape on the performance of sheet piles, Foundations: Proc. of the Second British Geotechnical Association Conference on Foundations, ICOF 2008, Vol. 1, pp. 319–330.

91) Taenaka, S., White, D. J. and Randolph, M. F. (2010): The effect of pile shape on the horizontal shaft stress during installation in sand, Proc. of the 7th International Conference on Physical Modelling in Geotechnics 2010, pp. 835–840.

Horizontal bearing characteristics of pressed-in piles

92) Li, Z. (2010): Piled foundations subjected to cyclic loads or earthquakes, PhD thesis, Cambridge University, 290p.
93) Kirkwood, P. (2015): Cyclic lateral loading of monopile foundations in sand, PhD thesis, University of Cambridge.

Group effect on vertical bearing capacity

67) White, D.J. (2002): The use of pressed-in H-piles for larger foundation structures, Proc. of the 7th BGA Young Geotechnical Engineers' Symposium, Vol. 1, 2p.
72) Deeks, A.D. (2004): An investigation into the strength and stiffness of jacked, driven and bored piles, M.Eng. Project Report, Cambridge University Department of Engineering, 51p.
94) Deeks, A.D., White, D.J. and Bolton, M.D. (2005): A comparison of jacked, driven and bored piles in sand, Proc. of the 16th International Conference on Soil Mechanics and Geotechnical Engineering, Vol. 4, pp. 2013–2106.
95) Yetginer, A.G., White, D.J. and Bolton, M.D. (2006): Field measurements of the stiffness of jacked piles and pile groups, Geotechnique, Vol. 56, No. 5, pp. 349–354.

Research topics to understand the function and the performance of embedded walls and structures

Performance of steel sheet pile walls

96) Kaneko, M. and Koseki, J. (2009): Analysis of bending strain characteristics of reinforced steel sheet piles used as coffer dam of an embankment of liquefiable ground, Proc. of 64th JSCE Annual Meeting, Vol. 64, pp. 647–648. (in Japanese).
97) Koseki, J., Tanaka, H., Otsushi, K., Nagao, N. and Kaneko, M. (2009): Model test of levee reinforced by sheet piles, Seisan Kenkyu, Institute of Industrial Science, the University of Tokyo, Vol. 61, No. 6, pp. 113–116. (in Japanese).
98) Kaneko, M., Tanaka, H., Otsushi, K., Nagao, N. and Koseki, J. (2010): Model test on the reinforcement of levees by the steel sheet piles, Part 3, Proc. of 65th JSCE Annual Meeting, pp. 563–564. (in Japanese).
99) Nagao, N., Tanaka, H., Otsushi, K., Kaneko, M. and Koseki, J. (2010): Model test on the reinforcement of levees by the steel sheet piles, Part 1, Proc. of 65th JSCE Annual Meeting, pp. 559–560. (in Japanese).
100) Otsushi, K., Tanaka, H., Nagao, N., Kaneko, M. and Koseki, J. (2010): Model test on the reinforcement of levees by the steel sheet piles, Part 2, Proc. of 65th JSCE Annual Meeting, pp. 561–562. (in Japanese).
101) Otsushi, K., Koseki, J., Kaneko, M., Tanaka, H. and Nagao, N. (2011): Experimental study on the reinforcement of levees by the steel sheet piles, JGS Journal, Vol. 6, No. 1, pp. 1–14. (in Japanese).
102) Boonsiri, I. and Kittiyodom, P. (2017): Feasibility study on using sheet pile as mitigation measure for road failure alongside of canal in Thailand, International Press-in Association Newsletter, Volume 2, Issue 1, pp. 15–17.

103) Nishioka, H. (2017): Development of sheet pile foundation, International Press-in Association Newsletter, Volume 2, Issue 4, pp. 13–16.

104) Otani, J. (2017): A new steel Sheet-Pile method for countermeasures against the settlement of embankment on soft ground -development of PFS method-, International Press-in Association Newsletter, Volume 2, Issue 3, pp. 8–10.

105) Poh, T.Y. (2017): Control measures for installation and removal of temporary earth retaining walls, International Press-in Association Newsletter, Volume 2, Issue 1, pp. 11–14.

106) Zhuang, Y. (2017): The effect of bottom stabilisation on sheet pile pit, M.Eng. Project Report, Cambridge University Department of Engineering, 46p.

107) Kasahara, K. Sanagawa, T. Nishioka, H. Sasaoka, R. and Nakata, Y. (2018): Seismic reinforcement for foundation utilizing Sheet Piles and soil improvement, Proc. of the First International Conference on Press-in Engineering 2018, Kochi, pp. 555–562.

108) Matsuzawa, K., Shirasaki, K., Konya, S. and Suzuki, Y. (2018): Example of construction of Sheet Pile walls using the cyclic auger method or anti-seismic reinforcement of railway embankment, Proc. of the First International Conference on Press-in Engineering 2018, Kochi, pp. 489–496.

109) Momono, K. and Fujita, M. (2018): Press-in with Augering; an installation of steel sheet piles connected longitudinally (hard ground press-in method), Proc. of the First International Conference on Press-in Engineering 2018, Kochi, pp. 453–458.

110) Nozaki, T. (2018): The press-in method with Augering - Augering area in relation to retaining wall design, Proc. of the First International Conference on Press-in Engineering 2018, Kochi, pp. 501–506.

111) Nozaki, T. (2018): Retaining wall deflection control in relation to Augering area, Proc. of the First International Conference on Press-in Engineering 2018, Kochi, pp. 507–516.

112) Shibata, N. (2018): A case study of design change in the press-in method, Proc. of the First International Conference on Press-in Engineering 2018, Kochi, pp. 467–474.

Performance of embedded wall against surcharge loading on retained soil

113) Maeda, T., Terai, T., Harada, F. and Motoi, Y. (2011): Design method of inclined-braceless retaining wall, Proc. of 66th Japan Society of Civil Engineers Annual Meeting, pp. 11–12. (in Japanese).

114) Nukui, K., Sakahira, Y., Imamura, M. and Hashizume, Y. (2011): Case study on inclined-braceless retaining wall for construction of water intake pipe on thermal power plant, Proc. of 66th Japan Society of Civil Engineers Annual Meeting, pp. 9–10. (in Japanese).

115) Aoki, S., Terai, T., Yoneya, S. and Yasunari, N. (203): Case study on inclined-braceless counterfort retaining wall method, Proc. of 68th Japan Society of Civil Engineers Annual Meeting, pp. 367–368. (in Japanese).

116) Aoki, S., Maeda, T., Moriyama, K. and Okawa, Y. (2014): Case study on inclined-braceless with ground anchor, Proc. of 69th Japan Society of Civil Engineers Annual Meeting, pp. 1405–1406. (in Japanese).

117) Gao, G. (2014): Comparing performance of different sheet pile walls, M.Eng. Project Report, Cambridge University Department of Engineering, 50p.

118) Ishihara, Y., Ogawa, N., Okada, K. and Kitamura, A. (2015): Implant preload wall: a novel self-retaining wall with high performance against backside surcharge, Proc. Of 5th IPA International Workshop, Press-in Engineering 2015, pp. 68–82.

119) Ogawa, N., Ishihara, Y. and Kitamura, A. (2017): Experimental study on deformation of selfretaining sheet pile wall due to excavation and backside surcharge, Journal of Japan Society of Civil Engineers, Ser. C (Geosphere Engineering), Vol. 73, Issue. 1, pp. 62–75. (in Japanese).

Embedded wall to reduce ground deformation associated with adjacent construction or increased embankment load

120) Research committee of PFS method (2005): Technical material on the partial floating sheet-pile method, 263p. (in Japanese).

121) Hizen, D., Kijima, N. and Ueno, K. (2018): Centrifuge model tests and image analyses of a levee with partial floating sheet-pile method, Proc. of the First International Conference on Press-in Engineering 2018, Kochi, pp. 215–220.

122) Imanishi, T. and Kajino, K. (2018): Case studies: use of the Gyropress method in tubular pile earth retaining walls for foundation works in urban area, Proc. of the First International Conference on Press-in Engineering 2018, Kochi, pp. 475–480.

123) Kashiwagi, K., Muto, S., Tobita, T. and Otani, J. (2018): Centrifuge modeling on the stability of a new steel sheet pile method during earthquake shaking, Proc. of the 53rd Japan National Conference on Geotechnical Engineering, pp. 1503–1504. (in Japanese).

124) Matsuzawa, K., Shirasaki, K., Konya, S. and Suzuki, Y. (2018): Example of construction of sheet pile walls for anti-seismic reinforcement of railway embankment, Proc. of the First International Conference on Press-in Engineering 2018, Kochi, pp. 481–488.

125) Nakai, K., Noda, T., Taenaka, S., Ishihara, Y. and Ogawa, N. (2018): Seismic assessment of steel sheet pile reinforcement effect on river embankment constructed on a soft clay ground, Proc. of the First International Conference on Press-in Engineering 2018, Kochi, pp. 221–226.

126) Tanaka, K., Kimizu, M., Otani, J. and Nakai, T. (2018): Evaluation of effectiveness of PFS method using 3D finite element method, Proc. of the First International Conference on Press-in Engineering 2018, Kochi, pp. 209–214.

127) Yamamoto, S., Kasama, K., Ohno, M. and Tanabe, Y. (2018): Seismic behaviour of the river embankment improved with the steel sheet piling method, Proc. of the First International Conference on Press-in Engineering 2018, Kochi, pp. 227–232.

Performance of pile walls embedded in hard ground

122) Imanishi, T. and Kajino, K. (2018): Case studies: use of the Gyropress method in tubular pile earth retaining walls for foundation works in urban area, Proc. of the First International Conference on Press-in Engineering 2018, Kochi, pp. 475–480.

128) Ishihama, Y., Fujiwara, K., Takemura, J. and Kunasegaram, V. (2018): Evaluation of deformation behaviour of self-standing retaining wall using large diameter steel pipe piles into hard ground, Proc. of the First International Conference on Press-in Engineering 2018, Kochi, pp. 153–158.

129) Kitamura, M. and Kamimura, S. (2018): Cantilevered road retaining wall constructed of 2,000 mm diameter steel tubular piles installed by the gyro press method with the GRB system, Proc. of the First International Conference on Press-in Engineering 2018, Kochi, pp. 445–452.

130) Kunasegaram, V., Takemura, J., Ishihama, Y. and Ishihara, Y. (2018): Stability of self-standing high stiffness-steel pipe sheet pile walls embedded in soft rocks, Proc. of the First International Conference on Press-in Engineering 2018, Kochi, pp. 143–152.

131) Kuroki, Y., Nishioka, H., Kumabe, K. and Koda, M. (2018): Reinforcement of damaged bridge pier by scouring using steel pipe piles, Proc. of the First International Conference on Press-in Engineering 2018, Kochi, pp. 549–554.

132) Miyanohara, T., Kurosawa, T., Harata, N., Kitamura, K., Suzuki, N. and Kajino, K. (2018): Overview of the self-standing and high stiffness tubular pile walls in Japan, Proc. of the First International Conference on Press-in Engineering 2018, Kochi, pp. 167–174.

Performance of embedded wall enclosure as countermeasures against liquefaction

133) Suzuki, K., Tei, K., Ohbo, N. and Hayashi, H. (1995): Evaluation of countermeasure against liquefaction using sheet-pile-ring, Soil Mechanics and Foundation Engineering, Vol. 43, No. 7, pp. 31–33. (in Japanese).

134) Sakemi, T., Tanaka, M. and Yuasa, Y. (1996): Settlement of oil storage tank during liquefaction, Doboku Gakkai Ronbunshu, No. 547, III-36, pp. 57–65. (in Japanese).

135) Matsuo, O., Okamura, M., Tsutsumi, T. and Saito, Y. (1998): Report on shaking table test on the construction method of sheet pile coffer dam as a liquefaction measure of embankment, Public Works Research Institute Report, 3539. (in Japanese).

136) Japanese Geotechnical Society (2004): Liquefaction countermeasure method, Maruzen Publishing Co., Ltd., 513p. (in Japanese).

137) Koda, Y. (2013): A study on design and construction for countermeasures against liquefaction by pressing steel sheet pile, M. Eng. thesis, Waseda University, 117p. (in Japanese).

138) Kato, I., Hamada, M., Higuchi, S., Kimura, H. and Kimura, Y. (2014): Effectiveness of the sheet pile wall against house subsidence and tilting induced by liquefaction, Journal of Japan Association for Earthquake Engineering, Vol. 14, Issue. 4, pp. 35–49. (in Japanese).

139) Noda, S., Kobori, Y., Sawada, K., Yashima, A., Fujiwara, K. and Otsushi, K. (2014): Study on reinforcement method of coastal levees against huge earthquakes, Part 1: model test on shaking table, Proc. of 49th Japanese Geotechnical Society Annual Meeting, pp. 1627–1628. (in Japanese).

140) Noda, S., Kobori, Y., Sawada, K., Yashima, A., Fujiwara, K. and Otsushi, K. (2014): Study on reinforcement method of coastal levees against huge earthquakes, Part 2: examination by numerical analysis, Proc. of 49th Japanese Geotechnical Society Annual Meeting, pp. 1629–1630. (in Japanese).

141) Ogawa, N., Ishihara, Y., Ono, K. and Hamada, M. (2018): A large-scale model experiment on the effect of sheet pile wall on reducing the damage of oil tank due to liquefaction, Proc. of the First International Conference on Press-in Engineering 2018, Kochi, pp. 193–202.

Performance of structures with piles as countermeasures against tsunami

142) Mitobe, Y., Adityawan, M.B., Ro, B., Tanaka, H., Otsushi, K. and Kurosawa, T. (2014): Hydraulic experiment on the effect of reinforcement on levees with double steel sheet pile walls against tsunami overflow, Proc. of 69th Japan Society of Civil Engineers Annual Meeting, pp. 61–62. (in Japanese).

143) Mitobe, Y., Otsushi, K., Kurosawa, T., Adityawan, M.B., Ro, B. and Tanaka, H. (2014): Experimental study on the effect of reinforcement on levees with steel sheet pile wall structures against tsunami overflow, Journal of Japan Society of Civil Engineers, Ser. B2 (Coastal Engineering), Vol. 70, No. 2, pp. I_976-I_980. (in Japanese).

144) Nakayama, T., Furuichi, H., Hara, T. and Nishi, T. (2014): Verification of stability of levee reinforced by double steel sheet piles against tsunami, Proc. of 49th Japanese Geotechnical Society Annual Meeting, pp. 975–976. (in Japanese).

145) Nakayama, T., Furuichi, H., Hara, T. and Nishi, T. (2014): Verification of performance of levee reinforced by continuous steel sheet pile/steel tubular pile walls against tsunami, Proc. of 69th Japan Society of Civil Engineers Annual Meeting, pp. 805–806. (in Japanese).

146) Oikawa, S., Kikuchi, Y., Kawabe, S., Mizuno, R., Moriyasu, S., Tanaka, R., and Takenaka, S. (2014): An experimental study for reinforcing construction of breakwater with structural steel, Proc. of the Special Symposium of Japanese Geotechnical Society on the 2011 Great East Japan Disaster, Tokyo, Japan, pp. 703–709. (in Japanese).

147) Arikawa, T., Oikawa, S., Moriyasu, S., Okada, K., Tanaka, R., Mizutani, T., Kikuchi, Y., Yahiro, A. and Shimosako, K. (2015): Stability of the breakwater with steel pipe piles under tsunami overflow, Technical note of the port and airport research institute, No. 1298. (in Japanese).

148) Furuichi, H., Hara, T., Tani, M., Nishi, T., Otsushi, K. and Toda, K. (2015): Study on reinforcement method of dykes by steel sheet-pile against earthquake and tsunami disasters, Japanese Geotechnical Journal, Vol. 10, No. 4, pp. 583–594. (in Japanese).

149) Ports and harbours Bureau (2015): Tsunami-Resistant Design Guideline for Breakwaters Reference V The technology conductive to Resilient Structures of the breakwaters by private sector businesses and others, pp. 3–20. (in Japanese).

150) Suguro, M., Kikuchi, Y., Hyodo, T., Kiko, M, Nagsawa, S., Moriyasu, S. and Oikawa, S. (2015): Horizontal loading experiment on breakwater reinforced by steel pipe piles, Journal of Japan Society of Civil Engineers, Sre. B3 (Ocean Engineering), Vol. 71, Issue. 2, pp. I_617-I_622. (in Japanese).

151) Dobrisan, A. (2016): Suitability of jacked-in steel piles as tsunami defences. M.Eng. Project Report, Cambridge University Department of Engineering, 48p.

152) Kubo, M. (2016): Construction in Okitagawa discharge channel, International Press-in Association Newsletter, Volume 1, Issue 1, pp. 6–8.

153) Suzuki, N., Ishihara, Y. and Isobe, M. (2016a): Experimental study on influence of porosity and material of pile-type porous tide barrier on its tsunami mitigation effect, Journal of Japan Society of Civil Engineers, Ser. B3 (Ocean Engineering), Vol. 72, No. 2, pp. I-491-I-496. (in Japanese).

154) Suzuki, N., Ishihara, Y. and Isobe, M. (2016): Experimental study on tsunami mitigation effect of breakwater with arrays of steel tubular piles, Journal of Social Safety Science, No. 29, 2016.11, pp. 7–14. (in Japanese).

155) Takada, H. (2016): Restoration and reconstruction of Kyu-Kitakami River, International Press-in Association Newsletter, Volume 1, Issue 1, pp. 3–5.

156) Dobrisan, A., Haigh, S.K. and Ishihara, Y. (2018): Evaluating the efficiency of jacked-in piles as tsunami defences, Proc. of the First International Conference on Press-in Engineering, pp. 289–296.

157) Ishihara, Y., Okada, K. and Hamada, M. (2018): Comparison of pile-type and gravity-type coastal levees in terms of resilience to tsunami, Proc. of the First International Conference on Press-in Engineering 2018, Kochi, pp. 251–256.

158) Shibata, D., Nakamura, S., Annoura, Y., Hoshino, M., Sasabe, T. and Yoshinaga, M. (2018): Earthquake and tsunami disaster preventive measures for sea embankment at USA fishing port, Kochi prefecture -application examples of press-in method to steel pipe pile installation, Proc. of the First International Conference on Press-in Engineering 2018, Kochi, pp. 523–532.

Researches on levees reinforced by double sheet pile walls

159) Tanaka, H., Tatsuta, M., Harada, N. Onda, K., Utsunomiya, S. and Nakano, K. (1999): Shaking table tests on earthquake resistance of river dikes with sheet pile walls driven at the top of slope, Proc. of 54th JSCE Annual Meeting, Vol. 54, 3(A), pp. 440–441. (in Japanese).

160) Tanaka, H., Tatsuta, M., Onda, K. and Utsunomiya, S. (2000): Seismic stability of river dikes with sheet pile walls driven at the top of slope, Proc. 35th JGS Annual Meeting, pp. 1563–1564. (in Japanese).

161) Okamura, M. and Matsuo, O. (2001): Dynamic centrifugal model test of double steel sheet pile temporary coffer dam on liquefiable ground, Proc. of 36th JGS Annual Meeting, pp. 1349–1350. (in Japanese).

162) Onda, K., Tatsuta, M., Tanaka, H., Saimura, Y. and Utsunomiya, S. (2001): Evaluation on a new liquefaction measure construction method by the steel sheet piles, Part 2: Evaluation by dynamic effective stress analysis, Proc. of 56th JSCE Annual Meeting, Vol. 56, 3(A), pp. 260–261. (in Japanese).

163) Tanaka, H., Tatsuta, M., Onda, K., Utsunomiya, S. and Saimura, Y. (2001): A study on a new liquefaction-countermeasure for embankments with steel sheet pile walls driven at the top of slope, Part 1: evaluation by shaking table model tests, Proc. of 56th JSCE Annual Meeting, Vol. 56, 3(A), pp. 258–259. (in Japanese).

164) Otsushi, K., Tanaka, H., Nagao, N., Fujiwara, K. and Koseki, J. (2011): Model test on the levee reinforcement using double floating sheet pile walls, Proc. of 66th JSCE Annual Meeting, Vol. 166, pp. 101–102. (in Japanese).

165) Suzuki, D., Iida, T., Ota, M., Fujioka, D., Yamamoto, T., Tanaka, H. and Otsushi, K. (2011): Fundamental study on aseismic reinforcement method of the river levees by the steel sheet piles, Proc. of 66th JSCE Annual Meeting, Vol. 66, pp. 109–110. (in Japanese).

166) Suzuki, D., Iida, T., Ota, M., Sumida, T., Tanaka, H. and Otsushi, K. (2012): Model test on aseismic reinforcement of river levee by the steel sheet piles, Part 1, Proc. of 67th JSCE Annual Meeting, pp. 601–602. (in Japanese).

167) Furuichi, H., Fukuchi, Y., Hara, T. and Otoshi, K. (2013): Technology development to mitigate long-term flood damages in the Kochi city due to the Nankai Earthquake, Nankai Earthquake and Disaster Prevention in the 21st Century, Vol. 7, Japan Society of Civil Engineers Shikoku Branch, pp. 43–50. (in Japanese).

168) Nakayama, T., Furuichi, H., Hara, T. and Otoshi, K. (2013): Verification of performance of double sheet pile levee against earthquake and tsunami by numerical analysis, Proc. of 48th Japanese Geotechnical Society Annual Meeting, pp. 1431–1432. (in Japanese).

169) Otsushi, K., Fruichi, H., Nishi, T. and Yoshihara, K. (2013): Dynamic effective stress analysis of levees reinforced by steel sheet piles, taking recovery of ground stiffness into consideration, Proc. of 10th Japan Earthquake Engineering Annual Meeting, pp. 299–300. (in Japanese).

170) Otsushi, K., Yoshihara, K., Fujiwara, K., Yasuoka, H. and Furuichi, H. (2013): Experimental study on disaster reduction technology of embankment structures with steel sheet piles, Proc. Nankai Geotechnical Symposium, Vol. 7, pp. 51–58. (in Japanese).

171) Otsushi, K., Yoshihara, K., Fujiwara, K., Yasuoka, H. and Furuichi, H. (2013): Verification of behavior of double steel sheet pile structure on a comparatively hard liquefiable ground, Proc. of 68th Japan Society of Civil Engineers Annual Meeting, pp. 5–6. (in Japanese).

172) Sumida, T., Iida, T., Ota, M., Otsushi, K., Hara, T. and Fujiwara, K. 2013. Model test on aseismic reinforcement of river levee by steel sheet piles, Proc. of 68th Japan Society of Civil Engineers Annual Meeting, pp. 1–2. (in Japanese).

173) Fujiwara, K., Sawada, K., Yashima, A., Abe, Y., Nakayama, H. and Otsushi, K. (2014): Experimental study on reinforcement measure of coastal levees by steel sheet piles under huge earthquake, Special Symposium on Getting over the Great East Japan Earthquake Disaster, JGS, pp. 417–423. (in Japanese).

174) Fujiwara, K., Yashima, A., Sawada, K., Abe, Y. and Otsushi, K. (2014): Analytical study on levees reinforced by double sheet piles with partition walls, 14th IACMAG International Conference, pp. 711–717.

175) Fujiwara, K. (2017): Reinforcement method for coastal dyke using double sheet-pile against large earthquake, PhD thesis, Gifu University, 123p. (in Japanese).

Others

176) American Petroleum Institute. (API) (2000): API Recommended Practice 2A-WSD: Planning, Designing and Constructing Fixed Offshore Platforms – Working Stress Design, 287p.

177) Finlay, T.C.R. (2001): Press-in piling: noise, vibration and the relief of hard driving. M.Eng. Project Report, Cambridge University Department of Engineering, 49p.

178) Japanese Geotechnical Society. (JGS) (2002): Deformation analysis of soils: from fundamental theory to application, p. 1. (in Japanese).

179) White, D.J., Finlay, T.C.R., Bolton, M.D. and Bearss, G. (2002): Press-in piling: ground vibration and noise during pile installation, Proc. of the International Deep Foundations Congress, Orlando, USA, ASCE Special Publication 116, pp. 363–371.

180) Rockhill, D. (2003): Ground vibrations due to construction operations. M.Eng. Project Report, Cambridge University Department of Engineering, 46p.

181) Rockhill, D.J., Bolton, M.D. and White, D.J. (2003): Ground-borne vibrations due to piling operations, Proc. of the International Conference organised by British Geotechnical Association, Vol. 1, pp. 743–756.

182) Construction Management Division, National Institute for Land and Infrastructure Management (2004): The new bid method promoted by the ministry of land, infrastructure, transport and tourism -the overall evaluation bid method, Striving towards to use skills and know-how on the public works, 8p. (in Japanese).

183) The United Nations Environment Programme (UNEP) (2004): c: Towards an Integrated Approach, 147p.

184) Boardmán, A.E., Greenberg, D.H., Vining, A.R. and Weimer, D.L. (2005): Cost-benefit analysis: concepts and practice (3rd Edition), Cambridge University Press, 576p.

185) Japan Society of Civil Engineers (2005): Challenge to introduce the asset management, GIHODO SHUPPAN Co., Ltd., 195p. (in Japanese).

186) Zhang, L.M., Ng, C.W.W., Chan, F., Pang, H.W. (2006): Termination criteria for jacked pile construction in weathered soils. J. Geotech. Geoenvironmental. Eng., ASCE, Vol. 132, No. 7, pp. 819–829.

187) Li, Z., Bolton, M.D. and Haigh, S.K. (2012): Cyclic axial behavior of piles and pile groups in sand, Canadian Geotechnical Journal, No. 49, pp. 1074–1087.

188) Tsukamoto, H. (2013): To be state of construction five principles of construction scientific assessment model for construction solution selection, Proc. of the 4th IPA International Workshop in Singapore, Press-in Engineering 2013, pp. 130–141.

189) Takeuchi, T. and Kimura, Y. (2015): Effective utilization of underground space in urban area, Proc. Of the 15th Asian Regional Conference on Soil Mechanics and Geotechnical Engineering, ESD-28.

190) International Press-in Association (IPA) (2016): Press-in retaining structures: a handbook, First edition 2016.

191) Ishihara, Y., Ogawa, N., Okada, K., Inomata, K., Yamane, T. and Kitamura, A. (2016): Model test and full-scale field test on vertical and horizontal resistance of hatted tubular pile, Proc. of the 3rd International Conference Geotec Hanoi 2016-Geotechnics for Sustainable Infrastructure Development, pp. 131–139.

192) Kitamura, A. (2017): Construction revolution–implant structure transforms global construction methods, DIAMOND Inc., 214p. (in Japanese).

108) Matsuzawa, K., Shirasaki, K., Konya, S. and Suzuki, Y. (2018): Example of construction of sheet pile walls using the cyclic auger method or anti-seismic reinforcement of railway embankment, Proc. of the First International Conference on Press-in Engineering 2018, Kochi, pp. 489–496.

193) Leung, C.F. and Goh, T.L. (2018): Noise and vibration monitoring for silent piling in Singapore, Proc. of the First International Conference on Press-in Engineering 2018, Kochi, pp. 579–586.

194) Yamaguchi, M. and Yamada, K. (2018): The press-in method assisted with Augering: case studies of single U and double Z shaped piles in the United Kingdom, Proc. of the First International Conference on Press-in Engineering 2018, Kochi, pp. 541–548.

195) Zeng, Y., Liu, C. and Zhang, L. (2018): Piling tests and induced surface settlement of rotating static pressure steel pipe pile in shanghai soft soil, Proc. of the First International Conference on Press-in Engineering 2018, Kochi, pp. 517–522.

196) Kitano, Y. (2019): Rehabilitation work following hurricane Katrina in New Orleans, USA, Press-in Piling Case History, Volume 1, pp. 119–120.

Index

Printed in the United States
by Bookmasters

Printed in the United States
By Bookmasters